The Endless Quest

The Endless Quest

Helping America's Farm Workers

Philip L. Martin
and David A. Martin

Routledge
Taylor & Francis Group

LONDON AND NEW YORK

First published 1994 by Westview Press, Inc.

Published 2019 by Routledge
52 Vanderbilt Avenue, New York, NY 10017
2 Park Square, Milton Park, Abingdon, Oxon OX14 4RN

Routledge is an imprint of the Taylor & Francis Group, an informa business

Library of Congress Cataloging-in-Publication Data
Martin, Philip L., 1949–
The endless quest : helping America's farm workers / by Philip L.
   Martin and David A. Martin.
      p.   cm.
   Includes bibliographical references and Index.
   ISBN 0-8133-1768-1
   1. Migrant agricultural laborers—Government policy—United
States.   I. Martin, David A., 1948–   . II. Title.
HD1525.M228   1994
331.5′44′0973—dc20   93-35889
                                                                             CIP

ISBN 13: 978-0-367-29176-1 (hbk)
ISBN 13: 978-0-367-30722-6 (pbk)

Contents

Boxes, Tables, and Figures

Figures

Foreword

A quarter of a century ago my wife, Jane, and I were rather strenuously involved in work with the children of migrant farm workers. We had gotten to know many black families who moved up and down the Atlantic seaboard states from May through October, harvesting an assortment of vegetables and fruit. We'd even met a few white families who did likewise – the last of a sizable number of such people who "followed the sun" (a phrase we'd heard used) during the 1920s and 1930s in order to make a living. Later, in the 1970s, when my family lived in New Mexico, I'd talk at great length in the Rio Grande Valley with Spanish-speaking children whose families also traveled long and far in pursuit of crops – up from Texas across the mountain states to the Pacific northwest. In 1967 I went with a crew from "the Valley" up to eastern Washington – and again, the sad spectacle of men, women, and (yes) children doing back-breaking work for the lowest of wages, under the most menial, even dangerous (pesticides) of circumstances. Eventually I would write up all that work, and in so doing, I had hoped that in time the "problem" of migratory farm labor would gradually disappear.

The government, back then, was actively involved in the matter; there was a United States Senate Farm Labor Subcommittee that periodically held hearings. Journalists, on television and in the newspapers, were anxious to document the appalling conditions that existed in various camps or indeed whole regions, such as the Lake Okeechobee part of Florida (Belle Glade, Bean City, Pahokee). The civil rights struggle had connected with that of migrants for a better life – so, surely, a nation newly attentive to discriminatory laws or customs would turn its attention to the way farm laborers were treated as they did their extremely important work.

Now, well into the nineties, only a few years from a new century, it is all too clear how naively optimistic some of us had been, as this thoughtful, carefully written book makes quite clear. Now, people from other countries more than make up for those migrants who have left the "stream," joined the rest of us who live and work in one place, belong to one or another neighborhood. The result is a continuing "pool" of hard-pressed, vulnerable, easily exploited labor; and the result, alas, is a sad story of sorts that won't go away.

Unfortunately for the many thousands of migrant farm workers who continue to work our American land, their plight is largely unknown, not documented with the kind of attention and concern needed. Hence the quite special value of this book – written out of great knowledge as well as an obvious moral concern. I can only hope, therefore, that this important and edifying effort will be rewarded by a wide readership, one that would surely be awakened by what is presented in the pages that follow. This is by far the most comprehensive and considered examination of this still vexing and all too melancholy subject to appear in print for some time, and many of us will surely hope that its arrival will be an occasion for yet another (and much needed) spell of political and educational responsiveness to the situation of those whose lives, whose daily, yearly condition is reflected in these pages with such careful regard.

Robert Coles
Harvard University

Acknowledgments

This book had its genesis in a charge to the National Commission on Migrant Education (NCME) to examine the degree of coordination between federal programs that serve migrants and seasonal farm workers (MSFWs) and their dependents. NCME asked the Administrative Conference of the United States (ACUS) to study these questions and generate initial recommendations for improvements. ACUS in turn asked us to produce the basic report, on the basis of field interviews with service providers, coupled with economic and demographic analysis and legal and historical research. Based on our report, ACUS in June 1992 recommended that the President should establish an Interagency Coordinating Council to strengthen national coordination among MSFW programs and that this Council should develop a system to obtain better data on migrant and seasonal farm workers. The ACUS recommendation is reprinted as Appendix A. NCME included its own recommendations for enhanced coordination in its final report, *Invisible Children: A Portrait of Migrant Education in the United States* (1992).

This book represents a revised and expanded version of our earlier report. Some of the more technical material has been eliminated here or shifted to an appendix, and other material, particularly Chapter 6 on the future of MSFW assistance programs, has been added. The views expressed here are the authors' and do not necessarily reflect those of the members of the National Commission, the Administrative Conference, or its committees.

A great many people helped us with this project. Jeffrey Lubbers, Research Director of ACUS, and Lisa Carlos, of the NCME staff, were especially insightful in helping to map our initial research strategy, arranging contacts with key people, providing a wealth of information, and suggesting improvements in the written product. Robert Suggs of the NCME staff also provided useful reports and data. Service providers at the local, state, and national levels and federal program administrators gave willingly of their time to explain their programs and wrestle with the issues we raised. We cannot list them all, but Patrick Hogan, of the Migrant Education Program, was especially generous with helpful suggestions and information, and Tony Fowler, also with ME, prepared the directory of federal MSFW service programs printed in Appendix B.

David A. Martin
Philip L. Martin

Introduction:
Sisyphus in the Fields

Migrant farm workers are perhaps the most pitied but least understood component of the American work force. Their plight has produced enduring literature, such as John Steinbeck's The Grapes of Wrath, and some of the best remembered television documentaries, such as Edward R. Murrow's Harvest of Shame. The position of farm workers near the bottom of the American labor market has been chronicled in numerous popular, scholarly, and government publications. Many have self-explanatory titles, such as Migrant and Seasonal Worker Powerlessness, A Caste of Despair, and The Slaves We Rent.[1]

The common image of a migrant farm worker depicts a hard-working Hispanic who lives during the winter months in southern Florida, southern Texas, or central California. Every spring, he packs up his family and follows the sun northward to harvest the ripening crops from New York to Michigan to Washington. Stories abound of migrants being abused by the unscrupulous farmers and farm labor contractors or crew bosses who employ them. While working, migrant farm workers are excluded from the protections of many federal labor laws, including those that cover collective bargaining and overtime pay, and governments at all levels have acknowledged their limited ability to enforce effectively other laws that are meant to protect migrants. Finally, "no large group of migrants has ever remained permanently migratory-- the best evidence that people are not migrants by choice."[2] As a result, society is often called upon to help migrant farm workers adjust to nonfarm lives and jobs.

The Endless Quest to Help Migrants

The Presidential platforms of both major political parties in 1960 called for action on the problems faced by migrant farm workers. The Democratic platform promised "to bring the 2 million men, women, and children who work for wages on the farms of the United States under the protection of existing labor and social legislation; and to assure migrant labor, perhaps the most underprivileged of all, of a comprehensive program to bring them not only decent wages but also an adequate standard of health, housing, social security protection, education, and welfare services." The Republican platform pledged an "improvement of job opportunities and working conditions of migratory workers."[3] These efforts, it was hoped, would push migrant farm workers up from their traditional place at the bottom of the American economy.

The federal government began programs in the mid-1960s to help migrant workers and their families. At that time, according to the best of chronically uncertain estimates, there were almost 500,000 U.S. citizen migrant workers, many of whom traveled across state lines to harvest crops. Federal assistance seemed necessary to overcome the reluctance of state and local governments to assist short-term migrant workers. Many local communities had no long-term interest in their well-being, wanting them to depart as soon as the harvest was over.[4] Federal programs for migrant workers and their families multiplied during the 1970s and 1980s, and today 12 programs spend over $600 million annually to assist migrant and seasonal farm workers (MSFWs) and their families (Box 1). Federal expenditures under these programs are equal to about 10 percent of what these workers earn, the equivalent of $600 to $700 per MSFW per year, but federal assistance efforts have been unable to nudge the status of migrant workers up the job ladder.

One reason why migrant farm workers and their families remain near the bottom is because federal programs to help migrant workers and their families often provide lasting aid by helping *individuals* to escape from the farm labor market. The migrants and their children who find nonfarm jobs often obtain higher incomes, but if they are replaced in agriculture by even more desperate workers, the programs that aid migrants call to mind the toils of Sisyphus in Greek legends. Because Sisyphus had angered Zeus, he was condemned forever to push a boulder uphill in Hades, only to see it roll back to the bottom each time. In America's fields, for any particular family that moves up the job ladder, assistance personnel turn around to see another migrant family in need of help taking their place.

Box 1. Migrant and Seasonal Farm Workers

There are about 2.5 million hired farm workers in the United States. Of these, about 2 million help to produce crops on the nation's farms; crop production includes more seasonal employment peaks and troughs than livestock production, and hence most MSFWs are employed on crop farms. About half of these 2 million crop workers are employed more than 1 month in agriculture, but less than 10 months, and they depend on seasonal farm work for most of their earnings. These workers earn an average $5000 annually for 6 months of work on farms, or $5 billion annually.

Throughout this book, it is most useful to think of the migrant farm workers who follow the crops from state to state--the original focus of the assistance programs we describe--as numbering about 300,000. These 300,000 migrants who follow the crops are accompanied by about 115,000 children. A larger number of workers--perhaps 600,000-- shuttle into the United States but then remain at one U.S. residence while they do farm work. They are accompanied by about 400,000 children. A few workers first shuttle into the United States and then follow the crops, for a total migrant work force of between 800,000 and 900,000. (Most federal MSFW assistance programs consider both follow-the-crop and shuttle workers to be "migrants.") These data are drawn from a number of sources, including the National Agricultural Worker Survey (NAWS) described in Chapters 2 and 5.

It should be emphasized that the data in this area are remarkably inadequate. Moreover, some federal MSFW programs serve more workers than are indicated by this description, such as the Migrant Education Program's inclusion of the children of year-round workers employed on livestock farms (if they moved within the last six years) and of nonfarm workers employed in high turnover food processing plants. Other MSFW programs serve fewer workers than are indicated by this description, such as the Job Training Partnership Act 402 program, which limits its services to workers legally authorized to be in the United States, are employed at least 25 days in agriculture, and who obtain at least 50 percent of their earnings from, or spend 50 percent of their working time on, farm work.

There is an alternative to using federal programs to ameliorate the deficiencies of the farm labor market--the farm labor market could be reformed so that migrants do not need special assistance programs. The U.S. government, however, has been unwilling to tackle the economic interests and the folk truths that are often used to justify the migrant labor system.

The most prominent myth is that poor farm workers are the price that the society must pay for cheap food. The facts disprove this myth. Two-thirds of the nation's farm work is done by farmers and their families, leaving only one-third to be done by hired workers. Migrant farm workers, the poorest hired workers, do less than half of the work done by hired workers. If there were no migrant farm workers, almost 90 percent of the nation's farm work would still be done. In other words, the migrant labor system that impoverishes hundreds of thousands of workers holds down the average family's food bill only a little. Even in the case of the fruits and vegetables that migrant workers often harvest, farm wages account for less than 10 percent of the retail price of a head of lettuce or a pound of apples. Doubling farm wages, and thus practically eliminating farm worker poverty, would raise retail food prices of even the crops picked by migrants by less than 10 percent.[5]

Instead of reforming the migrant labor system, the U.S. government has been persuaded to open the border gates to foreign farm workers. both documented and undocumented. These laborers, who usually have no other U.S. job options, have proved to be willing to accommodate themselves to the low wages and migratory lifestyles of seasonal agriculture. So long as foreigners without other U.S. job options are available to be migrant farm workers, however, little effective pressure exists to persuade farmers to improve wages and to eliminate migrancy. Thus a vicious circle is created: vulnerable migrant workers are available and so migrancy continues; migrant workers aspire to nonfarm jobs, and some of them escape; and the resulting farm labor shortages are used to justify the admission of more alien workers.

The United States today is still engaged in this endless quest to help migrant workers. This book reviews the federal programs that make often heroic efforts to push migrant farm workers up the U.S. job ladder. But so long as the federal government responds to familiar assertions, such as the plea of California farmers a century ago that "What we need is 40,000 to 50,000 good young Chinamen, "[6] the migrant labor system will not be changed—only the faces of the workers who need help.

What would it take to alter the migrant labor system so that special federal assistance programs are unnecessary? A century ago, a California observer laid out the problem, arguing (although with a somewhat racist cast common to such discussions in the 19th century) that it would be impossible to solve the farm labor problem without reforming the farm labor market:

> Hitherto the one great objection to an increase of the unskilled white labor population in California has been, that necessary as it was to have more help during the summer and harvest, the manner of husbandry in this state was such as to assure those who labor for others, work only for three, or at the highest, five or six months during the year. It was admitted to be an unnatural condition of affairs, and one which should be remedied, but which, under prevailing circumstances, could not be changed, especially as long as Chinamen in sufficient

numbers could be hired during the busiest seasons of the year...

The coming of Chinamen was tolerated and encouraged for many years. As a natural consequence they made for themselves a place in the industrial economy of the State, preventing thereby the natural increase and provision for a white laboring population. Employers could not expect white laborers to spring out of the ground when the Chinese influx ceased; nor can they now expect to remedy the evil, which a short-sighted policy... brought upon them, without suffering their consequences.

But the great danger is that they are unwilling to suffer these consequences and rather than undergo the annoyance of a settlement which would, once for all, put the questions of labor upon a right basis, they will look to the immediate future and continue to encourage or begin again to encourage Chinese immigration...If the size of their landed estates and the mode of cultivating them preclude the employment of civilized labor under civilized conditions, it is better that such estates lay waste, than that they be made a means of perpetuating the coolie system.[7]

The migrant labor problem has been understood for a long time. A 1951 Presidential Commission described the system through which the United States obtains migrant workers as one which "depends on misfortune to build up our force of migratory workers and, when the supply is low because there is not enough misfortune at home, we rely on misfortune abroad to replenish the supply."[8] The federal government's willingness to open side- or back-door border gates for migrant workers helped to produce wages and working conditions in agriculture that American workers will not tolerate. Immigration has thus helped to turn the farm labor market into a revolving door, through which workers without any other job options enter. Ten to fifteen years later, most leave the farm labor market because they are too old and slow to keep up with the younger and faster workers who continue to arrive.

Migrant farm workers are probably the largest needy work force in the United States. Although data are scanty, between 800,000 and 900,000, or about or 45 percent of the nation's 2 million crop workers, migrate a significant distance from their usual homes to help produce crops on the nation's farms each year. In most cases these migrant workers leave homes in rural Mexico and travel to a single farming community in the United States. Some migrants still follow the ripening crops around the United States with their families, but a lack of temporary housing for families has discouraged family migration.

Migrant workers are very diverse, but the average migrant worker is a 28 year-old Mexican-born male who earns $5 hourly or $5000 annually for 25 weeks of farm work.[9] Since manufacturing pays $10 hourly, and offers year-round work, many farm workers want to find factory jobs and thus quadruple their earnings.

The Plan of This Book

Few migrant workers find it easy to make the transition to a nonfarm job, and their children, whose lives may be disrupted by their parents' mobility, may also have trouble finding a nonfarm job. For these reasons, the federal government 30 years ago began special assistance programs to help migrant workers and their families. This book is primarily about the "big four" MSFW programs that help migrant workers and their families: the Migrant Education Program (ME), Migrant Head Start (MHS), Migrant Health (MH), and job training and ancillary services under section 402 of the Job Training Partnership Act (JTPA 402). These four programs together account for 90 percent of federal migrant assistance funds that serve migrant farm workers exclusively; ME alone counts for 56 percent.

This book is meant, initially, to shed light on those programs, by describing their evolution and their current operations. Chapter 1 reviews the role of migrant workers in U.S. agriculture, explaining that migrant workers are primarily associated with the one-sixth of the industry that produces fruits and vegetables. Chapter 2 then explains why each migrant assistance program was started, how it currently functions, whom the program serves (including attention to the detailed and disparate definitions that identify their target populations), and how it coordinates its efforts with other programs. No comparable book exists, providing even these basic descriptive facts on the assistance programs. The literature on these subjects is thin, and much of the information that has been gathered over the years is scattered through government agencies' or consultants' reports that are not easily accessible.

But our objective is not merely descriptive. We also offer three tiers of suggestions for reforms that would enable the assistance programs better to achieve their ultimate objectives. First, over the short term, better results could be obtained through improved coordination among the programs. We sketch out in some detail in Chapters 3 and 4 the current means used for coordination at the local, state, and national level, and the options available for improving national-level coordination. Ultimately we recommend the creation by statute or executive order of an interagency council with specific immediate tasks and a future mandate to consider wider reforms that transcend agency boundaries.

Second, over the medium term, all assistance to farm workers could be better targeted and better monitored if the data on the nation's farm workers were improved. Though numerous counts of migrant and seasonal farm workers exist, they paint widely differing pictures. This disparity makes it hard to measure the success or failure of existing programs, or to alter those programs to match changes in conditions or in farm worker activity.

New counts or estimates of the migrant population rarely serve as correctives for old ones, and cumulative improvements in data are rare – largely because of differences in methodology and differences in the definition of just what workers are to be counted in the relevant categories. Chapter 5 describes these problems and lays out a strategy, which would take many years to implement, to develop better data. Two elements are central: agreement on core elements of the definitions of migrant and seasonal farm worker, and clear assignment to a single agency of the responsibility to gather consistent farm labor data. We recommend the Bureau of Labor Statistics for this role – the agency that now produces the well-known monthly unemployment statistics and other labor data, but which has heretofore focused only on the nonagricultural work force.

Chapter 6 is our look into the future and it contains the third tier of our recommendations, covering a far longer time horizon. The migrant farm labor problem is not likely to be solved during the 1990s by some technological, social, or market revolution, such as mechanization, unionization, or the North American Free Trade Agreement (NAFTA). We argue that the federal government will have to take deliberate steps to improve conditions for migrant farm workers, and that the single most effective step would be to reduce the number of workers competing for farm jobs by better enforcing immigration and labor laws. The alternative of a continuing influx of desperate workers, and then low wages and poor conditions that prompt ever tougher laws and more extensive services, has failed. Enforcement and services are needed, but they will not solve problems in a labor market awash with workers.

The appendices pull together information that is often difficult to find, including the contact persons for each of the major programs (Appendix B) and the most recent farm labor data that can be arrayed on a state-by-state basis (Appendix C). Because this field has its own highly specialized jargon, we provide in Appendix D a glossary of the acronyms often employed. In addition, we have liberally used boxes to explain definitions, concepts, and issues that may not be understood by the diverse people interested in migrant workers.

In preparing this book, we reviewed the available literature and the legislative histories of the various programs, and we conducted interviews with those involved in administering migrant assistance programs at the federal, state, and local levels. Those interviewed included government officials, officers and employees of grantee service providers and of their umbrella organizations, farm worker advocacy organizations, and farmers and farm workers themselves. We also attended several hearings of the National Commission on Migrant Education.

At times this was a difficult study to conduct. Migrant assistance programs are flexible federal programs that serve mostly poor and minority workers and their families. The services are generally provided by hard-working and dedicated people, putting in long hours for pay that is often far below what they might command in other endeavors. The demand for farm worker services usually outstrips supply, leading some service providers to regard questions about whom they were serving as a hostile effort by outsiders that would only make the provider's daunting mission more difficult. In the face of an extensive need for services, our interest in definitions and efficiency could have appeared as uncharitable nit-picking, and some service providers thought that their expertise should simply be acknowledged to determine who should be served and how they should be served. However, most persons interviewed were gracious and freely devoted considerable amounts of time to assist us to understand their programs and to offer reflections on proposed reforms. We are indebted to them.

NOTES

1Migrant and Seasonal Worker Powerlessness, 1969-1970: Hearings Before the Subcomm. on Migratory Labor, 91st Cong., 1st and 2d Sess. (1969-1970)   (16 volumes of hearings produced under the direction of Senator Walter Mondale); Ronald L. Goldfarb, Migrant Farm Workers: A Caste of Despair (1981); Truman Moore., The Slaves We Rent (1965). See also Robert Coles, Migrants, Sharecroppers, Mountaineers (1971) (volume 2 of his classic "Children of Crisis" studies).

2Senate Subcomm. on Migratory Labor, The Migratory Farm Labor Problem in the United States, S. Rep. No. 1098, 97th Cong., 1st Sess. 1 (1961).

3Id. at x.

4The U.S. Senate noted that as of 1960, state residence requirements often barred migrants from all but emergency welfare assistance, so the Subcommittee recommended that "welfare assistance should be made available to the migratory farm family without regard to the question of residency." Id. at 23-24. According to the Subcommittee, 39 states required recipients of General Assistance to have lived in the state for 6 months to 3 years before applying for aid, and 40 states required residence for 1 to 5 of the preceding 9 years to be eligible for assistance under federal-state programs such as AFDC.

[5]This argument is spelled out in the supplemental views of Philip Martin, Report of the Comm'n on Agricultural Workers, 155-166 (1992).

[6]Quoted by Varden Fuller *in* Cal. Senate, Senate Fact Finding Comm. on Labor and Welfare, California's Farm Labor Problems: Part I, at 14 (1961).

[7]Varden Fuller, The Supply of Agricultural Labor as a Factor in the Evolution of Farm Organization in California, Exhibit 8762-A *in* Violations of Free Speech and Rights of Labor, pt. 54: Hearings Before the Senate Comm. on Education and Labor, 76th Cong. 3d Sess. 19777, 19815-19816 (1940) (quoting an 1884 statement by John S. Enos of the California Bureau of Labor Statistics)

[8]Presidential Comm'n on Migratory Labor, Migratory Labor in American Agriculture 3 (1951)

[9]A more complete description of the wages and working conditions of migrant workers is found in Richard Mines et. al., U.S. Farmworkers in the Post IRCA Period (Washington DC: U.S. Department of Labor, 1992).

1

Migrant Farm Workers in U.S. Agriculture

Migrant farm workers are persons who move in order to do farm work. A Presidential Commission defined a migrant as a "worker whose principal income is earned from temporary farm employment and who in the course of his year's work moves one or more times, often through several States."[1] This definition captures the familiar image of a person moving from farm to farm, and on each farm, doing work for two weeks to two months.

Other terms are often used to complete the picture: grueling work, low wages, few benefits, poor housing, child labor, and abusive employers.[2] Although there is no generally accepted definition of migrant farm worker, migrants, however defined, have been near the bottom of the U.S. job ladder for over a century, and their place there has been a concern for governments and organizations concerned about the well-being of all workers. The Presidential Commission noted in 1951 that there were migrants engaged in seasonal employment in nonfarm industries, but "only in agriculture [has] migratory labor become a problem of such proportions and complexity as to call for repeated investigations by public bodies."[3]

The U.S. Food and Fiber System

Agriculture is considered to be the oldest and, by some measures, the largest industry in the United States (Box 1.1). The farmers and farm workers who actually produce crops and livestock on farms are often described as the keystone of the larger food sector, which includes both the industries that provide farmers with fertilizer and equipment and the

Box 1.1.   Farming and Farm Employment

For purposes of the Census of Agriculture, a farm is "any place from which $1,000 or more of agricultural products were produced and sold or normally would have been sold during the census year," the years ending in 2 and 7. According to this definition, there were 2.1 million farms in 1991. About 20 percent of all farms are in Texas, Missouri, and Iowa. The largest 107,000 farms each sold farm products worth $250,000 or more, and they accounted for 56 percent of gross cash income. The smallest 1.3 million farms each sold farm products worth $20,000 or less, and they accounted for 5 percent of gross cash income. These small farms lost $500 million farming.

Farmers had cash marketing receipts of $170 billion in 1990, including 53 percent from the sale of livestock products and 47 percent from crop products. Production expenses were $144 billion, but with government payments and other farm cash income, net farm income was $51 billion. Exports of farm products in 1990 were $40 billion, or 11 percent of the total $366 billion in U.S. exports. U.S. imports of agricultural products were $23 billion, 5 percent of U.S. imports, leaving a net agricultural trade surplus of $17 billion.

The United States is a net exporter of the fruits and vegetables MSFWs pick. In 1990, U.S. farm exports included $3 billion worth of fruits and nuts (both fresh and processed), and $2 billion worth of vegetables (fresh and processed), up sharply from 1986 levels of $2 billion and $1 billion, respectively. Imports of fruits and nuts (fresh and processed) were worth $2.5 billion in 1990, and imports of vegetables were worth $2.3 billion. Tomatoes (fresh and processed) worth $400 million were the largest vegetable import.

About 4.6 million people lived on farms in 1990. However, most of the 2.5 million farm residents who are in the labor force are not employed in agriculture: 52 percent are employed off the farm. Most farm workers also live in towns and cities; although data are inadequate, it is believed that fewer than 10 percent of U.S. farm workers live on farms.

The U.S. Department of Agriculture and many farm organizations argue that farming is more important to the U.S. economy than employment and sales data would indicate. Employment on farms averaged 2.5 million in 1990, accounting for 2 percent of the nation's 117 million employed persons. Another 5.4 million workers were employed in industries that provide inputs to farmers – almost 3/4 of them provided farmers with services such as banking, research, and insurance. A further 13.2 American million were employed to manufacture and distribute food products, including 10 million employed in food retailing and restaurants. (Data drawn from the Census of Agriculture and USDA Economic Indicators of the Farm Sector.)

industries that process, transport, and distribute food to consumers. Most of the jobs and economic activity in the food sector is in the nonfarm economy: farming accounts for only 2 percent of U.S. jobs, and less than 2 percent of GDP.

The American farming system is often considered a crown jewel of the U.S. economy, a system which is envied around the world because it produces such an abundance of farm products that surpluses of food and fiber rather than shortages have been the major U.S. agricultural problem for over half a century. The United States is a net exporter of food and fiber, and the ability of the nation's farmers to produce more than Americans need has helped to hold down food prices. Americans devote only 12 percent of their expenditures for personal consumption to food and beverages, versus 15 to 20 percent in Western Europe, 42 percent in Japan, and over 50 percent in India.[4] The average American family spends about $28,400 annually, including $4300 on food. About $400 of these annual food expenditures are for the fresh fruits and vegetables that migrants often pick, roughly equivalent to what Americans spend on life insurance.[5]

Agriculture has long been a special industry. Producing or obtaining food is an essential task of all economic systems, and agriculture, which employed most Americans as recently as 70 years ago, is a special case in history, economics, and government regulation. The history of the United States is in large part the story of how an agrarian nation that believed that family farmers were the backbone of democracy was converted into an industrial nation. Several million farmers selling wheat or corn made agriculture a leading example of how competitive markets functioned, and the seemingly cruel fate of impersonal market forces leaving too many farmers and their families in poverty led to elaborate government programs to prop up farm incomes by establishing minimum guaranteed prices for the nation's major crops. The federal government intervened in product markets to ensure that family farmers got fair prices and thus fair incomes, but government often refused to intervene in the farm labor market under the theory that family farms should not be bothered with rules and regulations just to employ a hired hand who would soon be a farmer in his own right.

Agriculture has changed. Fewer and larger farms produce more of the nation's food and fiber; the largest 5 percent of all farms, each a significant business, account for over half of the nation's farm output, while the smallest two-thirds of the nation's farms account for 5 percent of all farm output. Farmers have become more integrated into the nonfarm economy: nonfarm businesses supply inputs to farmers that range from credit to chemicals, so that farmers are affected directly and indirectly by interest rate and pesticide regulation changes. Farmers have also become more integrated into the world economy: exports of farm products account for almost one-fourth of the value of U.S. farm production, so changes in the value of the dollar and trade policies can swing agriculture from boom to bust.

Public policies have tried to help farmers and their families buffeted by these changes in the national and global economies. Government policies continue to increase and stabilize farm incomes by assuring guaranteed prices for major field crops such as wheat, corn, and cotton. Government policies have also helped farmers to understand the scientific and economic factors that affect their biological factories by establishing land grant universities and extension services to educate them. But in comparison to these efforts to educate and assist farmers, government efforts to help farm workers represent a history of neglect.

Farm Labor: A History of Neglect

This neglect of farm workers can be attributed to a number of factors. Family farms—defined by the U.S. Department of Agriculture (USDA) as those that can operate with less than the equivalent of one and one-half year-round hired hands—have been the goal of the American farming system since colonial times. Agriculture's job ladder imagined "hired hands" to be temporary additions to the farm family who would soon become farmers in their own right; for this reason 19th century folklore often includes stories of youthful hired hands marrying farmers' daughters. When the farm population began to shrink after 1935, it was assumed that farm workers could best be helped after they left the farm, so there was more emphasis on maintaining full employment and regulating the nonfarm labor market than on intervening to improve conditions in the farm labor market.

As historians have noted, the neglect of farm workers was rooted in more than the assumption that most would soon become farmers or nonfarm workers. A permanent class of farm workers, especially in a country that offered free land to all who wanted to farm, was an uncomfortable topic in a nation founded on the ideals of equality. Slavery was the system initially used in the South, of course, to assure plantation agriculture a supply of labor to produce cotton, tobacco, and other crops for distant European markets. Farms in the western United States depended on hired farm workers, many of whom were immigrants who had no other U.S. job options, to help them produce crops for distant markets in the eastern states.

There was an explicit discussion in California of the need for farm workers who had no other U.S. job opportunities. When the *California Farmer* in 1854 asked "where shall the laborers be found...to become the working men" on the state's large farms, there was a wistful reference to slavery, but then an acknowledgment that "slavery cannot exist here." Nevertheless, farm work was described as " the work of the slave." The magazine repeated its question and offered a solution: "Then where shall the laborers be found? The Chinese! ...those great walls of China are to be broken down and that population are to be to California what the African has been to the South."[6] Furthermore, a California farm spokesman in 1872 observed that hiring seasonal Chinese workers who housed themselves and then "melted away" when they were not needed made them "more efficient ... than Negro labor in the South [because] it [Chinese labor] is only employed when actually needed, and is, therefore, less expensive" than slavery.[7] Neither slaves nor immigrant farm workers were topics which fit easily into discussions of how to bolster the family farm way of life that linked Americans to the ideals of their founding fathers.

Farm workers were recognized as a permanent feature of the U.S. farm system only in the 1960s, and even this belated recognition often assumed that many migrant and seasonal farm workers (MSFWs) would soon be displaced by machines. For example, a major three-volume study of labor in the fruit and vegetable industry concluded in 1970 that "the number of individuals who now do migratory farm work is now at an all time low of 257,000 and can be expected to decline still further."[8] This report echoed others in the 1960s by recommending "basic education, training and retraining programs for displaced adult farm workers," and justified this recommendation by asserting that "the door to employment is rapidly closing for those persons whose only qualifications for employment are a will to work and enough muscle to compete."[9]

The reports of the 1960s that recognized the existence of a permanent class of hired farm workers did so in the context of the revolutionary changes they expected in the farm labor market during the 1970s and 1980s. A 1960 magazine story captures the mood of the time: "The labor situation this year came to the forefront as the limiting economic factor in the state's agricultural development...within the next 5 years, agriculture in California is likely to see more of a shift to mechanization than it has in the past 20 years."[10] This story attributed grower demands for more mechanization research to the threatened end of the Bracero program (Box 1.2) and the union threat of the Agricultural Workers Organizing Committee. Although the focus was on reducing the need for 53,000 harvest workers--mostly Braceros--who were used in 1960 to harvest 132,000 acres of processing tomatoes, the story predicted that tree shakers and catching frames would soon be used to harvest cling peaches, apricots, nut crops, and "perhaps even cherries, apples, pears, etc." It was noted that mechanically-harvested commodities had to be handled in bulk, and that packers and processors were rapidly adopting bulk bins and forklifts to handle them even without mechanical harvesters in the fields, demonstrating the widespread belief that mechanization was inevitable.

Mechanization was expected to eliminate migrant farm labor and to improve the condition of farm workers. Mechanization, according to the story, has "one final advantage to which no monetary value can be placed...[A]s labor requirements are reduced, the need for ...the crew boss who manages the crew [disappears, leading to] the increased contact and closer relationship which the grower has with his employees."[11] There was little need for government worker assistance programs to reform the farm labor market because farmers were expected to make most of the on-farm adjustments necessary to develop the smaller trained work force they would need in a more mechanized agriculture. Farmers would also catch up with industry in recognizing "the relationship between a positive work environment and worker productivity and satisfaction." A positive work environment was defined as "adequate housing, timely and adequate transportation, field sanitation facilities, equipment, and mechanical aids to reduce physical strain."[12]

These expectations of fewer and better trained farm workers did not come to pass. Instead of a farmer-initiated reform effort in response to mechanization, and later the expected transitional role for federal farm worker assistance programs, the number of farm workers (migrants and non-migrants) began to stabilize in the 1970s at about 2.5 million.

The years between the mid-1960s and the mid-1970s were a golden era for many farm workers, marked by the end of the Bracero program in 1964, and the enactment of the nation's most pro-worker labor relations law in California in 1975. Union organizing and the general scarcity of farm labor raised farm wages sharply: California's average farm wage rose from $.95 per hour in 1965 to $2.43 in 1975, a 156 percent increase in 10 years. There were reports that a construction-style labor market was developing in at least parts of agriculture, marked by high wages when work was available, and unemployment insurance benefits when it was not.

Illegal immigration hastened the end of this golden era for farm workers. The number of aliens apprehended exceeded 1 million in 1977, and then rose sharply after 1982, until 1.8 million aliens were apprehended in 1986. Many of the aliens apprehended were Mexicans who were trying to find farm jobs in the United States. Today, 60 percent of all farm workers, and almost 90 percent of the migrant farm workers, are Hispanic.

Box 1.2. The Bracero Program: 1942-1964

The Bracero program refers to the series of agreements which permitted almost five million Mexican farm workers to enter the United States on a temporary basis to do farm work between 1942 and 1964. There were five million entries, but some workers returned year-after-year, so that only one to two million Mexicans participated. Bracero means day laborer. In Mexico, "Bracero" is often used to refer to "strong arms" in the same way that U.S. farm workers are sometimes termed hired hands.

A precursor to the Bracero program existed after World War I, when the U.S. government permitted 70,000 Mexican workers to enter the United States outside of normal immigration channels between 1917 and 1921. Mexican migration to the United States was then reduced to a trickle by Depression-era repatriations, but after the U.S. declaration of war on Japan and Germany late in 1941, southwestern farmers asked the U.S. government for permission to import temporary Mexican workers. The Mexican government was initially ambivalent about giving its blessing to a program which sent Mexican nationals to U.S. farms, where complaints of worker abuses were common. Mexico persuaded the U.S. government to guarantee the contracts issued to Braceros. In this way, the U.S. government became responsible for fulfilling any contractual provisions that U.S. farmers failed to satisfy.

The Bracero program was small during the war years, never exceeding 100,000 of the nation's 4 million hired farm workers. Braceros were concentrated on a few farms in a few states, and they worked in a handful of commodities. Fewer than 50,000, or 2 percent of the nation's commercial farms, ever employed Braceros. Although the major crop changed during the life of the program, most Braceros picked cotton during the program's first years.

When World War II ended, the Bracero program was expected to lapse. But growers argued that pre-war farm workers who had been drawn into factory jobs would not return to the seasonal farm work force, and that soldiers who had seen the world were also unlikely to be available to fill seasonal farm jobs. If the U.S. government wanted to produce as much food as possible to feed devastated Europe, farmers argued, Braceros would still be needed.

The Bracero program with the U.S. government as the contractor of Mexican workers nonetheless lapsed on December 31, 1947. There followed several years of informal and private U.S. employer recruitment of Mexican workers. Illegal immigration also increased as Braceros continued their established work patterns, but outside government-supervised channels. For many this meant learning that they did not have to pay bribes to local Mexican officials to get on recruitment lists, and then pay additional bribes at Mexican recruitment centers in order to work in the United States. U.S. farmers were pleased because they could employ Mexican workers without government "red tape," such as having their housing for Braceros iinspected and being required to offer them the minimum or

Box 1.2. The Bracero Program: 1942-1964

government-calculated prevailing wage, whichever was higher.

The informal postwar recruitment of Mexican workers sent Mexicans streaming north illegally. Many were legalized after they arrived in the United States, in a process that came to be termed, even in official U.S. government publications, "drying out the wetbacks."[12] The number of aliens who were legalized after arriving and finding employment illegally far exceeded the number of Mexican workers that U.S. employers contracted legally in the interior of Mexico. In 1949, for example, about 20,000 Mexicans received contracts from U.S. employers at recruitment centers in Mexico and legally entered the United States as contract workers, while over 87,000 arrived illegally in the United States and then had their status legalized after they found jobs.

A President's Commission on Migratory Labor was established to determine the effects of Braceros on U.S. workers. The President's Commission recommended that "no special measures be adopted to increase the number of alien contract workers beyond the number admitted in 1950," which was 67,500 Mexicans. The United States, it argued, could produce food for the Korean War emergency by using its domestic work force more effectively. Furthermore, the Commission recommended that "legislation be enacted making it unlawful to employ aliens illegally in the United States" and that "legalization for employment purposes of aliens illegally in the United States be discontinued and forbidden."[13]

These recommendations were not adopted. Growers tied their request for a new Bracero program to the nation's ability to win the Korean War, and in 1951, Congress enacted PL-78, the Mexican Farm Labor Program.[14] This is the program that is usually meant when describing "the" Bracero program. The number of Mexican workers in the United States rose sharply, reaching almost 500,000 in 1955.

The PL-78 Bracero program sowed the seeds for contemporary Mexico-to-U.S. migration by permitting U.S. agriculture to expand without raising wages significantly, thus creating jobs for seasonal workers that only rural Mexican workers were eager to fill. The Bracero program also added to supply-push factors by reinforcing patterns of development in rural Mexico which made millions of Mexicans dependent on the U.S. labor market. Finally, the Bracero program gave millions of Mexicans experience in the U.S. job market, so that after the Bracero program was terminated in 1964, millions of Mexicans knew U.S. employers who would hire them if they continued to arrive illegally. One commentator concluded that the Bracero program was responsible in part for the 1960s migrant, programs and protective legislation because Braceros had rights and privileges under the contracts growers were required to provide that U.S. farm workers did not have.[15]

Most migrant farm workers today are foreign born, having entered the United States either with or without documents. Migrant children are likewise mostly aliens, although a significant number of the children who accompany foreign-born migrants are U.S.-born citizen children. However, instead of following the ripening crops from state to state, most migrant children shuttle between U.S. and Mexican homes and schools. For example, a migrant child is more likely to attend one Mexican and one U.S. school each year than to attend two or more U.S. schools.[17] There are about 2.5 children who shuttle across the border for each migrant child who follows the crops.

The trend toward a shuttle immigrant farm work force should intensify in the 1990s. Over 90 percent of the new entrants to the farm work force are aliens, so that the alien share of the farm work force should rise above the current 60 percent, as retiring U.S. citizen workers are replaced by foreigners. U.S. agriculture should continue to employ about 2.5 million workers during the year in 2000, perhaps including by then 700,000 migrant workers who shuttle into and out of the United States. But the number who follow the crops within the United States is likely to shrink below its current level of about 300,000.

Federal migrant assistance programs have become the first points of contact for many of the aliens who are entering the U.S. labor market through agriculture. All indicators suggest that 1 to 2 million foreign citizens will continue to be employed in U.S. agriculture in 2000, making the effective delivery of education, training, health and related services to them a major challenge facing the United States.

Farm Labor Themes

Farm labor is usually considered a problem. A comprehensive report published by the California Senate in 1961 is entitled California's Farm Labor Problems. Six hearings around the state and several background studies produced a long list of farm labor problems, including the impacts of immigrant workers, problems with housing, and the enforcement of extant laws and regulations. The report did not recommend solutions or even prioritize the farm labor problems that it heard; it merely listed the different perspectives in a way which illustrates why it is so difficult to resolve them. For example, the California Farm Bureau Federation testified:

> there is an insufficient amount of capable domestic labor
> to meet the peak needs of California's agriculture.
> Because of this, farmers have been compelled to make
> use of such domestic labor as is available...[including

those who] are unemployable elsewhere. [Because farmers hire such workers] a problem of society as a whole has been placed on the farmers' doorstep.[18]

Farmers testified that they needed seasonal workers available at a reasonable cost on a varying schedule so that "California farmers can continue in the business of farming fruits and vegetables that are in competition with other parts of the United States."[19]

Worker advocates testified that farm workers needed "equality under the law" to overcome a long list of problems. Farm workers were described as "underprivileged, improperly fed, ill-housed, poorly clothed, inadequately socially protected...[and] poorly educated."[20] Seasonal workers were described as having problems finding enough jobs and earning enough to sustain themselves and their families. Farmers, worker advocates, government agency staff, and academics discussed both the general problems of farmers getting workers and farm workers finding work as well as specific problems, such as migrancy, housing, services for migrant children, illegal immigration, and technological changes that had been displacing and were expected to displace farmers and workers.

Farm labor problems are not new. Both the general and the specific problems of farmers, farm workers, and the agencies that serve workers and regulate the farm labor market have been discussed for over a century. Government has been urged to do both more and less about farm workers: advocates have often urged government to intervene and do more for farm workers, while farmers usually countered that more intervention was unwarranted.

These debates about what government should do often failed to define the farm labor problem that government was being asked to resolve. This is an important omission, since definitions of social problems usually include at least partial solutions. If the farm labor problem is defined to be low wages and earnings for the workers involved, then possible solutions range from higher minimum wages to tighter border controls which reduce the supply of workers. If, on the other hand, the farm labor problem is defined as how to have enough seasonal workers ready to work when weather and markets signal a farmer that it is time to harvest a crop, solutions range from having the government try to match available workers and jobs to importing temporary workers from abroad.

The proposed solutions have both equity and efficiency effects. For example, if the solution to the farm labor problem is to admit foreign workers, then farm employers may benefit but U.S. workers who compete with them may lose. Farm output may be larger and prices lower because the alien workers are available. There may also be second-round equity and efficiency issues, such as deciding who pays if immigrant workers settle and send their children to school. Defining and evaluating solutions to public policy problems can quickly become a complex exercise.

The persistence of farm labor problems despite a century of efforts to resolve them reveals that there are powerful incentives to maintain the status quo. The economic-political explanation for the persistence is straightforward. Farm wages have been lowered both by federal immigration policies, which let additional farm workers into the United States, and by labor laws which were slow to extend protections to farm workers. Just as farm prices that are raised by government policies increase land prices, so farm wages lowered by these government policies have been translated into higher land prices (benefiting farmers) and lower food prices (benefiting consumers). Changing governmental policies so that they raise rather than lower farm wages is difficult because the losers from higher farm wages resist such a change. A farm work force that includes a large proportion of immigrants without other U.S. job options does not usually have the economic power to persuade farmers to pay higher wages or the political power to change government policies.

The general farm labor problem persists in part because economic and political factors have maintained a power imbalance between employers and workers. Specific farm labor problems also persist. Many of the seasonal farm workers employed to produce fruits and vegetables are migrants, meaning that they move away from their usual homes and establish temporary residences to do farm work. These strangers-in-the-fields are often dependent on a local farmer or on a farm labor contractor for housing, work, and earnings. Migrancy gets farmers the seasonal workers they need, but at the cost of disrupted home lives and educational experiences for migrant workers and their families.

Public policy has been unable to resolve the migrancy problem. Despite repeated calls to reduce migrancy, public policies remain contradictory. On the one hand, public policies try to regulate in detail some aspects of migrancy in ways that may raise the costs of these non-local workers and thereby discourage the practice. They may, for example, specify exactly how migrant workers can be recruited and transported legally, or require that housing and other amenities be provided to migrant workers recruited through the federal-state employment system. At the same time, other public policies expend public funds to ease conditions for migrant and seasonal farm workers (MSFWs) and their families, perhaps lowering employers' costs and encouraging some workers to continue to migrate.

Why Migrant Assistance Programs?

The United States began as an agricultural nation, but for much of the past half century, farmers and farm workers could obtain higher wages, more benefits such as health insurance and vacation pay, and steadier employment by leaving the farm. The farm population peaked at 35 million in 1935, and is today less than 5 million. Sixty years ago, almost everyone who lived on a farm was a farmer or part of a farming family; today, less than half of those who live on farms are employed in agriculture. The average employment of hired workers has similarly declined, from perhaps 5 million, according to USDA's interpretation of the data (typically lower than other estimates) in the 1930s to about 2.5 million today. The rural-urban migration of farmers and farm workers peaked during the 1950s, when over 1 million Americans left the farm every year.

Migrant farm worker employment patterns are different. There were reportedly 2 million migrant farm workers in the 1920s, 1 million in the 1940s, 600,000 in the 1950s, and 400,000 in the 1960s.[21] During the 1970s, the number of migrant farm workers fell to 200,000. However, during the 1980s, estimated counts of migrant farm workers ranged from 100,000 to 1 million.

During the 1960s, several trends and concerns coincided to prompt the federal government to launch assistance programs to help migrant farm workers and their families to escape from agriculture. Even though over 800,000 Americans were annually switching from farm to nonfarm residences in the mid-1960s, agricultural economists argued that more of them would have to leave agriculture in order to create the relative labor shortages that were needed to raise farm wages and incomes.[22] Low farm wages, in their view, were due to "the redundant supply of labor in agriculture... agricultural labor is 'trapped' in the 'other America'."[23]

Congressional committees wanted to help migrant farm workers by enacting legislation to end the admission of Mexican Bracero farm workers, to eliminate child labor from the fields, and to regulate the farm labor contractors who matched many migrants with jobs. However, they were advised that it would be easier to have the federal government provide services for migrant farm workers than to regulate the labor market in which they worked. This service-instead-of-regulation strategy proved successful, and in September 1962 the federal government established Migrant Health, the first special assistance program for migrant farm workers.

Other federal programs for migrant workers and their families followed. In 1966, a Migrant Education Program was initiated to make grants to states to deal with the educational needs of the children of migrant farm workers. Job training for migrant and seasonal farm workers was provided under the Economic Opportunity Act of 1964, and the Office of Economic Opportunity that administered job training programs also provided support for Head Start programs that served the pre-school children of migrant farm workers.

The Federal programs for migrant farm workers that were launched in the mid-1960s were widely believed to be transitional programs, in the sense that they would be needed only until the mechanization of agriculture was completed in the mid-1970s. For these reasons, job training programs helped to train farm workers for nonfarm jobs, and the health and educational services provided to migrant children were meant to help them avoid becoming trapped in the migrant stream.

The expectation that migrant farm workers would disappear as had horse-drawn plows proved to be false. Although it is hard to measure precisely the number of migrant workers, the USDA reported that the number of migrant farm workers reached a peak of 466,000 in 1965, and then fell sharply to 159,000 in the last survey conducted in 1985.[24] The production and value of the fruit and vegetables that most of them picked, on the other hand, rose sharply.

Migrants and Immigrants

This picture of more production and fewer migrants is believed to be misleading, a reflection of flawed data rather than a window on reality. Some studies concluded that the number of migrant workers has held stable or even risen over the 1970s and 1980s. They surmise that the added migrants are invisible in USDA data because many migrants are Mexican nationals who are out of the United States when the surveys are conducted. Furthermore, migrants are difficult to enumerate while here because many are not legally authorized to be employed in the United States. A USDA study concluded that, based on production and other data, "migrants would logically have constituted an increasing, instead of declining, share of total hired labor...one explanation for this discrepancy is that foreign workers have accounted for an increasing share of migrant workers."[25] The belief that migrants were undercounted rather than displaced by machines seems to have been confirmed by events after the passage of the Immigration Reform and Control Act (IRCA) of 1986.[26]

IRCA was enacted to reduce the presence of unauthorized or illegal alien workers in the U.S. work force, largely by imposing sanctions on employers who knowingly hire unauthorized workers and by legalizing many of the undocumented residents already in the United States. IRCA sought to cushion agriculture's transition to a legal work force by including several provisions only for agriculture, such as the Special Agricultural Worker (SAW) program, with more generous terms for legalization than the general amnesty provisions,[27] the deferred enforcement of sanctions,[28] and a Replenishment Agricultural Worker (RAW) program that would admit probationary immigrant farm workers if farm labor shortages developed. [29]

Because the SAW program was a last-minute compromise that enabled IRCA to become law, predictions about the number and characteristics of the persons who would apply for legal status under its standards were highly speculative. One often-repeated expectation of how many unauthorized aliens were employed in U.S. agriculture, usually as migrant farm workers, was the USDA estimate that 350,000 illegal aliens were employed in agriculture in the early 1980s, and this number became the ceiling for Group I SAWs.[30] However, the major surprise of the SAW program was that 1.3 million aliens applied for SAW status, or over half of the estimated total hired farm work force of 2.5 million in the mid-1980s, and almost ten times the USDA estimate of *migrant* farm workers.

IRCA opened a new window on migrant farm workers. In order to determine if IRCA produced farm labor shortages, a new survey of farm workers, the National Agricultural Workers Survey (NAWS), was launched. In the course of finding that no farm labor shortages resulted, this Department of Labor-sponsored survey obtained demographic data on the workers employed in most of U.S. crop agriculture. The NAWS defined migrant workers as those who traveled 75 or more miles from their usual residences in search of farm work, and it found that 42 percent of the workers interviewed in the early 1990s made such a move. Compared to non-migrant workers, the migrants interviewed in the NAWS are more likely to be male and Hispanic, and twice as likely *not* to have their families with them at their farm worksites. [31]

The justification for migrant assistance programs and the characteristics of migrant farm workers have changed remarkably over the past 25 years as the workforce has undergone dramatic changes. What began as programs meant to help U.S. citizens who were trapped in the migrant stream to escape successfully into the nonfarm labor market have become, in many instances, integration programs for new immigrants, both documented and undocumented, who are finding their way into the U.S. economy through agriculture. The next chapter recounts how each of these programs evolved.

NOTES

[1]Presidential Comm'n on Migratory Labor, Migratory Labor in American Agriculture 1 (1951).

[2]Many distinguished Americans have studied migrant workers and their children. Robert Coles devoted one of his children of crisis books to *Migrants, sharecroppers, mountaineers.* (Boston: Little, Brown 1971); he had previously done a study entitled *The migrant farmer: a psychiatric study.* (Atlanta : Southern Regional Council, 1965).

[3]Presidential Comm'n on Migratory Labor *in* American Agriculture (Washington, DC, 1 (1951).

[4]Statistical Abstract of the United States, 833 (1992) (data are for 1988).

[5]Id. at 444 (data are for 1990, and they refer to expenditures by the nation's 97 million consumer units).

[6]Quoted *in* Varden Fuller, Hired Hands in California's Farm Fields 7 (1991).

[7]Quoted *in* Varden Fuller, The Supply of Agricultural Labor as a Factor in the Evolution of Farm Organization in California, Exhibit 8762-A, *in* Violations of Free Speech and Rights of Labor, pt. 54: Agricultural Labor in California, 19809, Jan. 13 Hearings Before the Senate Comm. on Education and Labor, 76th Cong. 3d Sess. (1940). The merits of slavery versus seasonal farm workers were debated extensively, with seasonal workers usually found to be cheaper because: (1) no capital outlay was required to purchase them; (2) they boarded themselves while employed and reproduced abroad; (3) they were available when needed but they were paid only for the time they were actually employed; and (4) at the end of the season they "moved on, relieving [the] employer of any burden or responsibility for his [worker's] welfare during the slack season." Id. at 19824.

[8]Fruit and Vegetable Harvest Mechanization 18 (B. F. Cargill and G. E. Rossmiller eds., 1970).

[9]Id. at 22.

[10]Dan Henley, Labor Troubles Speed Automation in California, Better Farming Methods, Dec., 1960, at 14-15.

[11]Id. at 32.

[12]Cargill and Rossmiller, supra note 7, at 19.

[13]Presidential Comm'n on Migratory Labor, supra note 1, at 53.

[14]Id. at 178-180.

[15]Pub. L. No. 78, 65 Stat. 119 (1951).

[16]Richard B. Craig, The Bracero Program 200 (1971).

[17]For a more detailed discussion of this shuttle migration, see the Chapter 5 review of the data developed by the National Agricultural Workers Survey (NAWS), from which many of the statistics in this section are drawn.

[18]Cal. Senate Fact Finding Comm. on Labor and Welfare, California's Farm Labor Problems: Part I, at 9 (1961).

[19]Id. at 10.

[20]Id.

[21]The Encyclopedia Britannica article on Migrant Labour provides these migrant estimates for earlier decades, although it cautions that "there are not many reliable statistics." 12 Encyclopedia Britannica 176 (15th edition, 1974).

[22]C.E. Bishop noted in 1966 that there was a net transfer of 25 million Americans from farm to nonfarm residences between 1940 and 1964, but, "in spite of this vast migration, the return for labor in farming remains comparatively low." C.E. Bishop, Dimensions of the Farm Labor Problem, *in* Farm Labor in the U.S. 5 (C.E. Bishop, ed., 1967).

[23]Id. at 13.

[24]Victor J. Oliveira, Trends in the Hired Farm Work Force, 1945-1987, at 5 (USDA, ERS, Agric. Info. Bull. 561, 1989).

[25]Id.

[26]Pub. L No. 99-603, 100 Stat. 3359 (1986).

[27]8 U.S.C. §1160 (1988).

[28]8 U.S.C. §1324a(i) (3) (1988).

[29]8 U.S.C. §1161 (1988).

[30] Group I SAWs did at least 90 days of seasonal agricultural services (SAS) work in each of the years ending in May 1, 1984, 1985, and 1986. Group II SAWs, by contrast, did 90 days of SAS work only in the year ending May 1, 1986. (Over 90 percent of all SAW applicants were in the Group II category.) The practical differences in treatment between the two groups were relatively minor. Both were fully entitled to "temporary resident" status, including work authorization, but Group I SAWs were allowed to become full permanent residents one year earlier than Group II SAWs.

[31]*Executive Summary* to Department of Labor, Findings from the National Agricultural Workers Survey 1990, at ii (1991).

2

The Development of Migrant Assistance Programs

Most materials on migrant farm workers reinforce the stereotype that migrants are minority families who are strangers-in-the-fields at their temporary workplaces and who have special needs and problems that are not addressed by their employers or by local assistance programs in the communities where they temporarily reside. Although this stereotype masks some of the diversity in today's farm labor force,[1] it has been the source of sympathy and federal initiative to address the migrants' needs. In many cases, that initiative has eventually led legislators or administrators to expand their programs so that they may also meet the needs of nonmigrant seasonal farm workers.[2] By 1992, federal expenditures for service programs intended specifically for migrant and seasonal farm workers (MSFWs) totaled over $600 million annually, equivalent to about 10 percent of migrant earnings.[3] (See Table 2.1.)

Overcoming Local Neglect

During the 1960s the federal government launched a War on Poverty, enacting or expanding numerous statutory programs for assisting poor and disadvantaged Americans. Some of these programs technically included farm workers within their compass, but it was often argued that migrant farm workers would be excluded or underserved by these state-administered programs. As a result, the government also established during the 1960s numerous programs specifically dedicated to serving migrant farm workers and their families.

Specially targeted federal efforts for migrants were justified on
several grounds. First, migrants have special needs that result simply
from the fact of their mobility. They may not be in one location long
enough to do the paperwork or to satisfy the waiting period for benefits
administered by state and local governments. For migrant children in
particular, frequent moves cause obvious disruptions in schooling.[4] The
resulting educational deficiencies might trap them in a culture of
migrancy and poverty. In the 1960s, it was feared that migrant children
would be unable to follow their parents into farm work because of
mechanization, but their educational deficiencies would also leave them
unprepared to compete in the nonfarm labor market. The withering
away of migrancy of course has not come to pass, but other changes have
strengthened the case for special assistance programs -- including the
problem of language barriers as the migrant work force has come to be
dominated increasingly by recent immigrants and U.S. citizens of
Hispanic origin.

Second, migrants (like Indians, who were increasingly the objects of
special subprograms) were often thought to be a unique responsibility of
the federal government.[5]   Migrants usually do not vote in the
jurisdictions in which they work (if in fact they have voting rights at all;
as foreign nationals, many cannot vote in the United States), and they are
often regarded by farming communities as a necessary evil needed to get
the crops picked. Migrants were described as people who "pass through
community after community, but they neither claim the community as
home nor does the community claim them."[6] Some communities feared
that making benefits available to migrant workers and their families
might encourage them to stay in the area, absorbing resources that, some
local leaders believed, should be reserved for residents who were
disadvantaged. Even without such local suspicion or hostility, there may
be little contact between migrants and permanent residents that would
lead to initiatives to help meet the migrants' needs. Local efforts
therefore could not be relied upon for assistance, and even the general
assistance programs funded by the federal government, and usually
administered by state or local officials through cooperative
arrangements, might not take adequate account of the particular needs of
migrants.

Special programs for migrants often mean that they may receive
benefits under programs that do not serve the nonmigrant poor -- or that
special funding is available to agencies serving migrants that they would
not receive if they served other segments of America's disadvantaged
population. There is nothing improper or surprising about this fact; it is
an inevitable corollary of the nation's recognition of special migrant
needs that require unique programs. But this difference does create
certain difficulties unperceived in the heady days of the early War on

Table 2.1. Federal Migrant and Seasonal Farm Worker Programs

Program	Department	Services	Funds ($Mil in FY92)	Funds ($Mil in FY88)	Percent Change
Migrant Education (ME)	Education	Funds state educational agencies (SEAs) to serve the children of migrants who are 3 to 21	$308.3	$269.0	14.6
Migrant Health (MH)	HHS	Funds clinics that provide primary health care for MSFWs and their dependents	57.7	43.5	32.6
Job Training Partnership Act 402 (JTPA 402)	Labor	Employment and training services for MSFWs and their dependents	77.6	65.6	18.3
Migrant Head Start (MHS)	HHS	Early childhood program for migrant children age 0 to 5	85.9	40.5	112.1
Total "Big 4" Programs			$529.5	$418.6	26.5
High-School Equivalency Program (HEP)	Education	Funds colleges and universities to assist migrants and their dependents to get a High School Diploma or equivalent	8.3	7.3	13.7
College Assistance Migrant Program (CAMP)	Education	Funds colleges and universities to assist migrants and their dependents to ease their transition into college	2.3	1.3	76.9

Program	Agency	Description			
Migrant Even Start	Education	Funds programs to coordinate child and adult education for migrants	2.1		
Migrant vocational rehabilitation	Education	Funds programs for handicapped migrants	1.0	1.1	-9.1
Women, Infants, and Children (WIC)	USDA	Provides food and nutrition counseling	17.5	13.0	34.6
Migrant legal services	LSC	Provides legal services to MSFWs	10.8	9.4	14.9
Section 516 MSFW housing grants	USDA	Grants to nonprofit organizations for farm worker housing	11.0	11.2	-1.8
Section 514 MSFW housing loans	USDA	Loans to farmers and nonprofits for farm worker housing	16.3	11.4	42.9
Community Services Blocks Grants	HHS	Block grant funds reserved for farm workers	3.0	3.0	0.0
Total			$601.8	$476.3	26.3%

MSFWs also participate in other programs for which they qualify, including Food Stamps, AFDC, literacy programs, homeless programs, bilingual and immigrant education, and low income home energy assistance.

Source: AFOP Washington Newsline, November/December 1991, at p. 3 and June 1988, at p. 3, supplemented by interviews of selected agency officials.

Poverty, when the nation appeared to have the capacity to address the problems of all of the disadvantaged. As budgets tightened and benefits declined in the 1980s, special and comparatively better services for migrants posed more acutely these questions: Who qualifies for such services? What priorities should be established to parcel out limited migrant assistance budgets? And how can the programs be managed or designed to minimize any possible incentive to manipulate migrant status in order to claim such benefits or expand bureaucratic turf?

For most programs today, these questions are answered by the eligibility provisions of each migrant assistance program -- often coupled with a first-come first-served rationing system for benefits that may run out well before all those eligible are served. But these definitions and eligibility provisions were enacted piecemeal; they reflect no overarching congressional or administrative theory about those most likely to benefit from, or most deserving of, the particular assistance at issue.

Each federal MSFW assistance program has a unique definition of the migrant and seasonal workers who are eligible for services. These definitions differ, for example, in the border which must be crossed to be considered a migrant, in the type and amount of qualifying work that must be done to be considered a migrant farm worker, and in how long a migrant can continue to receive services after he or she has stopped migrating. These differences in definition mean that each MSFW assistance program has a unique target population, distinct outreach workers and intake forms, and usually separate facilities that may be able to serve some farm workers but not others.

Moreover, migrant assistance programs have developed as a series of ad hoc initiatives run by different agencies. The result is a clear potential for overlaps in some services and gaps in others, and Congress has made inadequate provision for coordination between programs. Service providers generally concur that today's mix of programs and eligibility standards would not be replicated if the federal government were starting from scratch to develop an overall strategy to meet migrants' special needs. But, service providers are also fearful of a wholesale redesign of the programs in a era of tight budgets and a changing target population.

Migrant Education

The Migrant Education Program (ME or MEP) is the single most costly migrant assistance program; its $300 million accounts for over half of all federal funds designated for migrant workers and their dependents. ME is unusual in several respects. First, funds are made available to each state based on that state's share of the number of eligible children identified -- but not necessarily served -- by recruiters in that state. Second, ME supports its own data system, the Migrant Student Record Transfer System (MSRTS), to expedite the transfer of the school records of the children who are identified as eligible for ME services; the MSRTS also serves as the data base for state-by-state funding allocations. Third, ME services are usually delivered by public employees in school facilities; most other migrant assistance services are provided by the employees of private non-profit organizations. Finally, ME has expanded the criteria for eligibility more than other migrant programs. While other programs focus on currently migratory parents and children, ME serves mostly the children of migrants whose parents no longer cross school district lines in order to do farm work. ME also considers as eligible for services the children of more nonfarm workers employed in food processing industries than any other program.

In the heyday of the War on Poverty, Congress passed the Elementary and Secondary Education Act of 1965, establishing a significant federal role in supporting education.[7] Title I of that Act set a basic funding pattern that has been continued to this day for the major portion of federal assistance. Grants are made, through the states, for aid to local education agencies (LEAs) on the basis of counts of school-aged children from low-income families, largely based on decennial census figures. Once the funding is provided, however, in observance of the traditional local control over education in this country, LEAs have considerable discretion in choosing exactly how to use those funds, so long as they are used for supplemental services and facilities and not to fund the basic educational services of the school.[8]

The 1965 statute was reorganized and revamped by the Education Consolidation and Improvement Act of 1981 (ECIA), and then by the Elementary and Secondary School Improvement Amendments of 1988,[9] but the federal funding approach for educating the disadvantaged remains essentially the same under Chapter I, as it is now known. LEAs with a high concentration of children from low-income families receive supplemental federal funds that can be used, for example, to employ additional teachers and aides, to provide student counseling and tutoring, to support in-service training for Chapter I personnel, and for other activities, including advocacy for needy children. Chapter I funding of basic grants to LEAs for school year 1991-92 amounted to $5.0 billion.[10]

In 1966 Congress determined that an additional special program for migrant education was needed, based in part on concern that too much of the basic program's funds were going to urban areas.[11] Unlike basic Chapter I, the statute governing migrant education places primary responsibility on state educational agencies (SEAs), rather than local school districts, "[b]ecause of the transient nature of the population."[12] Migrant education grants, now known as Section 1201 grants, totaled $285.6 million for FY 1991, up from $274.0 million in FY 1990. This money was provided to the SEAs in the 49 states that now participate (all but Hawaii), plus the District of Columbia, Puerto Rico, and the Northern Marianas.

SEAs have considerable discretion in structuring actual services and deciding on the precise uses of the funds.[13] They may pay for counselors, tutoring, additional aides, dropout prevention programs, prevocational training, medical and dental services, nutritional programs, transportation, training or counseling of parents, special summer schools (a particularly important element in upstream states, during some of the principal fieldwork months for the parents, in order to assist students to make up work missed over the regular school year), and a host of other initiatives. This great flexibility is a notable feature of the ME program. Moreover, the money comes as an addition to basic Chapter I funding; one person termed ME money the "supplement to all other supplements."

In addition to the 1201 grants, section 1203 of the statute[14] provides for additional grants "to improve the interstate and intrastate coordination among State and local educational agencies of the educational programs available for migratory students." The FY 1991 appropriation for these purposes was approximately $9 million. The bulk of this money (some $ 6 million) supports the Migrant Student Record Transfer System (MSRTS), discussed below, while the rest is used for grants and contracts that promote coordination, including modest funding for services to migrant children at stopover sites and a new program to develop a better system for secondary education credit exchange and accrual.[15]

The key element for coordination, however, has been a system of three Program Coordination Centers (PCCs), located in Texas, Oregon, and New York, funded at the level of approximately $2 million annually.[16] Each center has a staff of four or five professionals, and they run workshops, provide training, share curricular materials, and furnish other services aimed especially at helping states understand the curricula employed in other systems where their migrant students may spend part of the year, so as to mesh more effectively the educational programs of migrant students. The ME office in Washington oversees this coordination through its Office of Program Coordination, staffed with approximately seven professionals. Most of the coordination activity of the office focuses on coordination among ME programs, but the staff also has responsibility for coordination with the other migrant service programs administered by other federal agencies.

As with other migrant assistance programs, some coordination among ME programs and with other agencies also derives from the work of umbrella organizations for ME grantees, especially the Interstate Migrant Education Council (IMEC) and the National Association of State Directors of Migrant Education (NASDME). IMEC is a 51-person organization of chief state school officers, ME directors, and federal, state, and local elected officials of the 16 states that include about 80 percent of all migrant children. Founded in 1976, it has been chaired since then by Representative William Ford (D-MI), Chairman of the House Education and Labor Committee, and widely regarded as the "father of ME." IMEC's purpose is to "disseminate information about the unique benefits of the migrant education program and the need for sufficient resources to carry out its mission."[17] Several persons also described IMEC as a device for raising the visibility of Migrant Education within SEAs. IMEC pursues these goals by educating public officials about the needs of migrant children, and by providing a forum for ME directors to interact with these officials. IMEC has a small staff which carries out the directives of the 32-member Council of elected officials and other non-ME program members and a 19-member steering committee of ME-affiliated staff. IMEC publishes a newsletter, *Migrant Education Report*, and other occasional papers. It is funded by SEA contributions of $22,000 to $25,000 per state; its budget in 1992 was $400,000.[18]

NASDME is the second private coordination organization. NASDME is an unincorporated organization of state ME directors whose chair rotates among the states. NASDME meets quarterly so that state ME directors can exchange information and promote coordination between ME programs in various states. State ME programs pay dues to NASDME to fund its operations, but data on these dues and NASDME's budget were not available to us. NASDME publishes a bimonthly periodical, *Migrant Education Messages and Outlook (MEMO)*, and hosts an annual conference. NASDME has devoted considerable attention to the MSRTS, focusing on strategies to improve the delivery of timely, reliable and germane information. It has also tackled the problems involved in coordinating ME with other programs, and it has worked on school credit transfers and related issues.

Which Children Are Eligible?

The Migrant Education Program serves children who move with or to join parents or guardians who move across school district lines "to obtain temporary or seasonal employment in an agricultural or fishing activity."[19] All children who participate in the MEP must have a valid

Certificate of Eligibility (COE) which explains the child's qualifying move. Until 1988, there were three ways in which agricultural or fishery activities could be considered temporary or seasonal, and thus the children of an adult employed in such an activity could be considered eligible for MEP services: (1) if the job or work was indeed temporary, such as harvesting apples; (2) if the farm worker asserted that he or she would be employed only temporarily, such as a worker who says she will milk cows for only three months; or (3) if the farmer reported that he was hiring the worker only temporarily, such as a farmer who says he is hiring a worker who moved into the area for only three months to milk cows.

In a 1988 change to the MEP policy manual, a fourth definition of temporary or seasonal was added which greatly expands the number of children eligible for MEP services. If a parent or guardian today crosses school district lines in order to work in an agricultural or fishery activity, even if (1) the activity is year-round, such as meat or poultry processing, (2) the employer asserts that the worker is being hired permanently, and (3) the worker says that she moved into the area to work permanently in the year-round job, the children are eligible for MEP services if the establishment has a worker turnover rate of at least 60 percent annually over 18 months.

This policy change more than doubles the pool of potentially eligible children. The food manufacturing industry (Standard Industrial Classification 20) employed an average 1.7 million persons in 1990, including 250,000 in preserved fruits and vegetables and 425,000 in meat processing plants. Since worker turnover in many food manufacturing industries is quite high, the number of persons employed sometime during the year is considerably higher than 1.7 million, perhaps twice average employment. A major effect of this policy change is to enable midwestern MEPs to enroll the children who move with their parents into the area's high turnover meat-packing plants. Many of the new entrants into the midwestern meat packing plants are Asian and Hispanic immigrants.

Taken to its logical conclusion, this policy could one day permit MEP to expand to encompass virtually the entire food and fiber sector, which includes, according to USDA, almost 17 percent of the American work

---

Box 2.1. Definitions: Migrant Education

20 U.S.C.A. §2782(c) (West 1990):

The Secretary shall continue to use the definitions of "agricultural activity", "currently migratory child", and "fishing activity" which were published in the Federal Register on April 30, 1985, in regulations prescribed under section 555(b) of the Education Consolidation and Improvement Act of 1981 and subpart 1 of part B of title I of the Elementary and Secondary Education Act of 1965 (as in effect on April 30, 1985). No additional definition of "migratory agricultural worker" or "migratory fisherman" may be applied to the provisions of this subpart.

34 C.F.R. §201.3(b) (1992):

Agricultural activity means:

(1) Any activity directly related to the production or processing of crops, dairy products, poultry, or livestock for initial commercial sale or as a principal means of personal subsistence;

(2) Any activity directly related to the cultivation or harvesting of trees; or

(3) Any activity directly related to fish farms

...

Currently migratory child means a child:

(1) Whose parent or guardian is a migratory agricultural worker or a migratory fisher; and

(2) Who has moved within the past 12 months from one school district to another--or, in a State that is comprised of a single school district, has moved from one school administrative area to another--to enable the child, the child's guardian, or a member of the child's immediate family to obtain temporary or seasonal employment in an agricultural or fishing activity. This definition includes a child who has been eligible to be served under the requirements in the preceding sentence and who, without the parent or guardian, has continued to migrate annually to enable him or her to secure temporary or seasonal employment in an agricultural or fishing activity. This definition also includes children of migratory fishermen, if those children reside in a school district of more than 18,000 square miles and migrate a distance of 20 miles or more to temporary residences to engage in fishing activity.

...

Fishing activity means any activity directly related to the catching or processing of fish or shellfish for initial commercial sale or as a principal means of personal subsistence.

Formerly migratory child means a child who:

(1) Was eligible to be counted and served as a currently migratory child within the past five years, but is not now a currently migratory child; and

(2) Has the concurrence of his or her parent or guardian to continue to be considered a migratory child.

Migratory agricultural worker means a person who has moved within the past 12 months from one school district to another--or, in a State that is comprised of a single school district, has moved from one school administrative area to another--to enable him or her to obtain temporary or seasonal employment in an agricultural activity (including dairy work).

Migratory children means children who qualify under either the definition of "currently migratory child" or "formerly migratory child" described in this section.

Migratory fisher means a person who has moved within the past 12 months from one school district to another--or, in a State that is comprised of a single school district, has moved from one school administrative area to another--to enable him or her to obtain temporary or seasonal employment in a fishing activity.

force, or some 21 million workers, although there is no indication that the MEP is planning such an expansion.

In 1966–67, the initial statutory provision for Migrant Education was read to cover children only in the year following their parents' migration. As a result, migrant children whose families settled out of the migrant stream could no longer be served after 12 months of settlement. State migrant education directors approached Congress soon after the MEP was launched in 1966 and asked for authority to continue MEP services for the children of parents who no longer migrated. They pointed out that the statute perversely terminated the assistance at a point when it might hold special promise of effectiveness, just when the student's life was taking on somewhat greater stability. Congress responded promptly, amending the statute in 1967 to extend the period of eligibility by another five years, if the parents concurred.[20] The legislative record discloses no particular reason for choosing the five-year benchmark rather than some other time period.[21]

The current statute and regulations maintain this approach, allowing services to "formerly migrant children" (sometimes simply called "formerlies") with the concurrence of the parents, for no more than five years.[22] In essence, this provision authorizes services for a total of six years after a "qualifying move," for one year as a "currently" and thereafter as a "formerly." The statute specifies, however, that currently migratory children "shall be given priority in the consideration of programs and activities" -- a command that is implemented in different ways in different states.[23] In some states there is little practical distinction between the services available to migrant children regardless of when their parents last migrated.

Apart from the lengthy "look-back" period MEP employs to determine if a child is eligible for services (the longest of all federal migrant programs), the MEP definition of qualifying work is also among the more expansive.[24] It covers the children of parents employed in both crop and livestock agriculture, and also, by special statutory provision, those employed in dairy and fishery operations. The definition also extends to the children of migrant workers in packing and processing facilities, as well as the children of some persons who transport agricultural products. A qualifying move need only take one across a school district line, and the worker need not actually find farm work in the new destination, as long as the move was undertaken with the *purpose* of finding agricultural work. This highly subjective element, requiring inquiry into the intent of the family at the time of the move, whatever the actual employment before or after, is open to manipulation. It sometimes causes problems for outreach workers, and it certainly complicates audits.

Allocating MEP Funds to States

States must apply to the federal Department of Education for § 1201 MEP grants, and their applications must demonstrate compliance with certain criteria, such as assuring adequate evaluation, making an assessment of current needs, and including ample provision for parental involvement. Nevertheless, the actual funding each SEA receives derives in nearly automatic fashion from a formula based on the identified migrant student population in the state during the year. The statute specifies proportional counting of eligible children present in the state for only part of the year, in order to yield a full-time-equivalent (FTE) migrant student population count for each state each year.

It is hard to calculate the number of migrant children and their location throughout the year because, as Chapter 5 explains, the underlying data on farm workers have always been weak and there is no reliable way to track the children. After some initial difficulties shortly after enactment of the program in 1966, the Office of Education decided to use Department of Labor (DOL) estimates of the number of migrant farm workers in each state. Even though it was aware of significant shortcomings in these data, at the time the office could find no better starting point for applying the statutory command. It then developed an estimate of the number of migrant school children per state by assuming that each worker was accompanied by 0.75 children.[25]

State directors of migrant education in 1967 appointed a committee to develop a system to transfer the records of migrant students from one school district to another, and in 1969, the state of Arkansas received a contract from the Office of Education to implement the system, which became the Migrant Student Record Transfer System (MSRTS), based in Little Rock. By 1972, all active states were cooperating in the use of this system, and the state directors came to realize that MSRTS could also be used to calculate each state's share of migrant children. In 1974, amendments to Title 1 required, in essence, that the MSRTS be used to allocate ME funds to states, and it has been used for this purpose every year since then.[26]

In 1974, the MSRTS reported that 162,000 migrant children were logged into its system. By 1990-91 some 628,150 children, almost 4 times as many, were reported by states to the MSRTS. During the 1980s, the number of formerly migrant children rose much faster than the number of currently migratory children, rising from 232,000 to 330,000, a 42 percent increase. Formerly migrant children also became more concentrated in the top five states, which had 70 percent of all such children in 1990-91. The number of currently migratory children, by contrast, rose only 17 percent, from 254,000 to 298,000, and the top 5 states saw their share of currentlies fall from 76 to 70 percent (Table 1).

Since MSRTS misses some of the nation's migratory children, the true number deemed eligible for the MEP might be more than 700,000.

There is a great deal of skepticism that the number of MSFWs' children has almost quadrupled while the number of farm workers was decreasing. However, there is no other nationwide data system that provides information on migrant children. This means that the validity of MSRTS data can be checked only for internal consistency and against data bases established for purposes other than enumerating migrant children. Not all MSRTS data are internally consistent. There are no systematic quality checks on the data submitted to MSRTS by State Educational Agencies (SEAs), so data entry errors as well as improperly identified children may be logged. SEAs must submit age and migratory status data to MSRTS to receive MEP funding, and many observers have noted that, since MEP funds are allocated to states on the basis of the number of eligible children the state has identified, states have an incentive to over-report. Finally, the MSRTS includes data only on the children that recruiters locate, so it cannot provide a complete profile of the nation's migrant children.[27]

The MSRTS was originally developed to expedite the sharing of academic records among school districts as migrant students moved from place to place with their families, not to provide a census count of eligible children. The MSRTS system has encountered problems and serious criticisms,[28] but once it was fully operational, it provided far better data on migrant *students* than the DOL estimates of migrant workers–hence its usefulness as the basis for the state-by-state allocations mandated by statute.

The formula for using MSRTS data to fund SEAs is quite complex, but the basic idea is that the federal supplement for each full-time-equivalent migrant child identified in the state will amount to 40 percent of the state's average per-pupil expenditure. This calculation is subject, however, to both a ceiling and a floor: it cannot fall below 80 percent, nor exceed 120 percent, of the national average per-child supplement.[29] For program year 1991-92 the statutory formula, had it been directly applied, would have generated federal grants to SEAs totaling $987 million. But because Congress appropriated only $286 million, the Office of Migrant Education first performed the statutory calculations, then reduced each state's allocation proportionately.[30] Although the formula itself (including the 40-percent benchmark) is not based on a precise assessment of educational requirements, these proportionate reductions to stay within appropriations ceilings lead to an oft-heard complaint that SEAs receive less than a third of the funding the statute says they need.

Current Issues

Since the MEP claims the largest portion of the federal migrant
assistance budget, more questions and complaints were raised about it in
the general community of MSFW service providers interviewed for this
study than any other program. These questions and complaints most
often voiced are summarized here.

Funding. Each state's share of MEP funds is based on the MSRTS
count of FTE eligible children within state boundaries, without direct
reference to how many of that population actually receive services. In
program year 1990, for example, when the MSRTS reported that over
556,000 children were eligible for services, a Migrant Education Fact
Sheet reported that approximately 250,000 children were served.[31]
Other reports, however, appear to establish that the number of children
served exceeded 350,000.[32] However, this 350,000 figure may double-
count children who had access to services in more than one state. The
MEP clearly serves only a fraction of the children that are deemed
eligible for services.

Funding-for-counting was the number one complaint about the MEP
program. Other programs, it was pointed out, usually have performance
standards that include linking requests for funds closely to the number
of persons served. For example, one Migrant Head Start official
observed that: "The principal use of MSRTS frequently appears to be the
identification of children to generate dollars for migrant education rather
than the utilization of a system to work more effectively with children to
promote their educational development and health maintenance."[33]

ME officials defend the formula, arguing that it provides an
incentive for finding and identifying migrant children. As a result, most
state or local ME programs have significant recruitment and outreach
staffs. Proponents of funding-for-counting argue that services are
generally made available once any significant concentration of migrant
students is identified, and they point out that MEP services certainly will
not be provided if the students are not located. And if they had
sufficient funds, they argue, they could serve all eligible children.[34]

Critics also complain about the way in which the MSRTS system
generates FTE counts. Even if MSRTS records show that a student has
withdrawn from a school system, she remains within that state's count
until she is picked up by a system in a different state and that state
notifies MSRTS of the enrollment there.[35] Supporters defend this
practice, however, based on a judgment that the student may simply
have dropped out of school, could still be in the area, and should be the
object of outreach efforts to draw her back in the MEP. If she has
actually moved elsewhere, she cannot remain on the rolls for longer than

about 12 months in any case, because MEP regulations[36] require the annual updating of information on the certificate of eligibility. No one knows exactly how many migrant children who leave one MEP and are not picked up in another are true dropouts; it is alleged that the child in some instances moves to another state, enrolls there in a MEP, but does not show up in the MSRTS immediately because the new MEP was slow to complete the paperwork or failed to note the student's status as a migrant.

Moreover, this slow removal from MSRTS counts may at one time have worked to the special advantage of stopover states. Several people asserted that Arkansas, in particular, provides an attractive stopover site at Hope, on the Texas border, for migrants moving north from Texas, and is assiduous in enrolling migrant children with the Little Rock-based MSRTS during their brief stay there (usually limited to a maximum of 12 hours). Most migrants reportedly arrive at night and leave early in the morning. Migrant children whose parents stopped there on the spring trip north were provided with educational services such as books and study plans, registered in MSRTS, and thus counted for funding by Arkansas for the rest of the summer if the children did not enroll in another MEP program in Ohio, Michigan, or another northern state. This practice generated disproportionate FTE counts for Arkansas, because some of the migrant children were not picked up in another state system, and the reporting of others' moves to MSRTS was delayed by weeks or months. In its latest regulations, however, the Department took steps to minimize this practice by stopover states.[37]

Other states also have policies that tend to increase the count of children eligible for services but do not necessarily increase educational services for migrant children. For example, the states of Washington, Oregon, and Arizona provide free accident and life insurance, whose premiums are paid with Federal MEP funds, to induce families to register their children for MEP services and thus have them included in the MSRTS. Each state pays premiums of about $200,000 per year, and claims are processed by state-level MSRTS personnel.[38]

"Formerlies" vs. "Currentlies." Many people interviewed questioned ME's lengthy "look-back" period, allowing children to be served by MEP for up to six years after their parents made a qualifying move. ME officials, teachers, and supporters usually justify the practice by pointing out that the effects of educational disruptions may linger for years. In their experience, there is no significant difference in the educational needs of formerlies and currentlies in classrooms or in counseling or tutoring sessions.[39] Moreover, some MEP officials assert that a shortening of the child's eligibility period might only tempt some families to rejoin the migrant stream in order to restart their children's eligibility for ME services.

Critics concede that a child's needs may persist long after a family settles out of the migrant stream, but they suggest that at some point, well before the six-year mark, the needs of settled ex-migrant children are not significantly different from those of other disadvantaged students. Basic Chapter I funding exists to provide services for sedentary disadvantaged children.[40] Critics argue that a reduced eligibility period would not provide an incentive to renew migrancy, because from a settled-out family's perspective, MEP services usually are not distinguishable from other services available through the local school system or elsewhere.

At the time when it was adopted in 1967, the statutory change that expanded coverage to the children of formerly migrant workers held no implications for funding, because state allocations then were based on DOL estimates of the number of active migrant farm workers in each state each month. The expansion of eligibility at that time allowed only a continuation of services, not an increase in some states' funding. But when the statute changed in 1974 to base state allocations on the MSRTS count of all eligible children in the state, the 1967 amendment to include "formerlies" took on a funding significance that has never been fully considered in Congress.

The fact that a majority of MEP students are now formerlies may justify Congressional re-examination of the 1967 amendment. California, which receives over one-third of all MEP funds, has more formerlies than currentlies.[41] Some of these formerlies, moreover, never did fit the stereotypical image of follow-the-crop migrants. Their ranks include, for example, Southeast Asian refugees who moved but one time (after their initial resettlement in the United States) to central California to take up truck farming on small plots owned by extended family members. That one move was "qualifying"; it fits the ME definition, for it was undertaken "to enable [a family member] to obtain temporary or seasonal employment in an agricultural ... activity."[42] Though they plan no further movement, their children then bring California added MEP funding for the next six years.[43] Again without doubting the existence of bona fide educational need on the part of these children, critics question whether their inclusion on migrant rolls, skewing funding to the heavy advantage of California -- and other similarly situated states -- provides for the best use of the limited MEP funds, especially when other special federal programs exist for the specific purpose of assisting the education of refugee children.

During interviews for this study, we found some slight indications of a willingness on the part of ME officials in Washington to consider a reduction in the look-back period, possibly to as little as 24 months, in the interests of moving toward a more uniform definition of who is a migrant farm worker. But most people involved in ME programs stoutly defend the option to provide services for a full five years after a qualifying move.[44]

A middle path on the issue of how long a child should be eligible for MEP services may be worthy of consideration. If MEP returned to the pre-1974 situation, which was arguably more in keeping with the initial purposes behind special federal programs for migrant education, funding could be based on a state count which includes only current migrants, or current migrants plus those in the first year of settling out. But the statute could still allow states, at their option, to include formerly migrant students in the services. Alternatively, the FTE formula could be weighted to provide only partial credit, for funding purposes, for formerlies, while giving full weight to the count of currently migratory children. None of these alternative systems would require significant changes in MSRTS data-gathering, for each student is already registered as a currently or a formerly. (On the COE and related forms, Status 1 and 2 are reserved for currently migratory agricultural workers, while all formerly migratory agricultural workers and their families are recorded in Status 3.) Any such change would of course bring significant political fall-out, for it would clearly shift resources from states like California and Texas toward those where a currently migratory work force is more prevalent.

Age Ranges. Until 1988, the MEP funded state programs on the basis of the number of children 5 to 17. In 1988, the statute was amended to include migrant children aged 3 to 21 for funding purposes. The House report that accompanied the amendment explained:

> Currently, States may serve children between the ages of 3 and 21, but only those migrant children between the ages of 5 and 17 are counted for funding purposes. By expanding the age range being counted for funding purposes, the committee hopes to draw attention to the need to correlate to some degree those children who are served with the amount of funding provided.[45]

Ironically, the change may instead have reduced the correlation between services and funding. The earlier range, from 5 to 17, corresponded fairly closely with the usual ages of local school attendance in grades K-12, in most districts. Although service was authorized for younger and older children, it was not the norm. Moreover, SEAs often do not need to create new programs targeted at such young children or older youths in order to include them in their counts. The eligibility documents filled out by MEP recruiters are supposed to show the names and ages of other members of the family, whether or not they are in school. Siblings 3 to 5 or 17 to 21 can now generate added federal funding, whether or not they are served and whether or not they were located through additional targeted outreach.

The expanded age range also compounds the problems of overlap with Migrant Head Start, which traditionally serves children from shortly after birth through age 5. One MHS provider interviewed, who operates a program that must turn away substantial numbers of eligible preschoolers for lack of resources, expressed some bitterness at what he saw as the indifference of the local ME office to pre-school children. Despite the MEP's wider mandate, he complained, ME would not think of transferring its funds to enable MHS to serve a larger population. The two were like "separate empires." When the local ME head was asked what his office was doing to serve the younger children, in view of the new statutory provisions, he mentioned only a summer school to help prepare those who would be entering kindergarten the coming fall -- that is, children already 5 years old or approaching that age. This implies no wrongdoing on the part of ME; by all appearances the ME monies provided to the district were effectively used, but they were focused on children 5 and older. The episode simply underscores the potential mismatch of the population counted for funding purposes with the population actually served. (It should be noted that, in some other states and areas, ME and MHS have worked together to more effectively to integrate their services.[46])

Coordination. The ME statute contains two specific provisions relating to coordination. Section 1203, discussed above, encourages coordination among states' ME programs and between state MEPs and LEAs.[47] The other provision, Section 1202(a)(2), allows the Secretary to approve a state ME application only if he determines, *inter alia*, "that in planning and carrying out programs and projects there has been and will be appropriate coordination with [the HEP/CAMP programs described below], . . . the Education of the Handicapped Act, the Community Services Block Grant Act, the Head Start program, the migrant health program, and all other appropriate programs under the Departments of Education, Labor, and Agriculture."[48]

This measure dates back to 1966, when it referred to migrant programs authorized under the Economic Opportunity Act, but it has gradually been expanded to reflect the reorganization of migrant programs and a wider range of programs with which coordination must be accomplished. No one interviewed contested the desirability of such coordination. But some ME officials bemoaned the absence of parallel requirements in the organic statutes governing the other migrant service programs, fearing that the other agencies, lacking similar statutory requirements, would not have the same interest that MEP has in coordination. This concern is probably exaggerated, however, for other programs have incorporated coordination requirements in their performance standards.

HEP and CAMP

The Office of Migrant Education in the Department of Education also oversees other programs meant for migrant farm workers or their families. The most important are the High School Equivalency Program (HEP) and the College Assistance Migrant Program (CAMP), which fund projects designed to help migrant students complete their secondary schooling and their first year of college, respectively.[49] These programs were originally established as discretionary grant programs of the Office of Economic Opportunity (OEO), HEP in 1967 and CAMP in 1972.[50] In 1973 the director of OEO delegated responsibility for the programs to the Secretary of Labor, and in 1978 Congress expressly established authority for the programs under § 303(c)(2) of the Comprehensive Employment and Training Act (CETA).[51] Their administration was transferred in 1980 to the newly created Department of Education, and later that year authority for the programs was removed from CETA and incorporated into the Higher Education Act of 1965, as amended.[52] FY 1990 appropriations for HEP totaled $7.9 million and for CAMP $1.7 million.[53]

HEP provides grants to colleges and universities, or private nonprofit organizations that work in cooperation with a college or university, to help MSFWs or their dependents (at least 17 years of age) to obtain a high-school equivalency diploma. HEP funds may be used for outreach to and the recruitment of eligible beneficiaries, for educational services, and to provide a wide range of supportive services, including counseling, health services, room and board (most HEP students live on college campuses during the program), and weekly stipends for personal expenses. Grants are awarded on a three-year cycle, through a competitive process that is not directly tied to migrant population in the area -- thus sharply distinguishing the program from basic ME § 1201. As of November 1990 there were 23 HEP grantees, located in 16 states and Puerto Rico.[54]

CAMP operates quite similarly, based on a competitive grant process with a three-year grant cycle. It is meant to assist eligible MSFWs and their dependents to begin college studies. Funds may be used for outreach and recruitment aimed at such persons "who meet the minimum qualifications for attendance at a college or university." Thereafter, during the first year of college, funds may be used for

---

Box 2.2. Definitions: HEP/CAMP

20 U.S.C.A. §1070d-2 (West Supp. 1993):

(b) Services provided by high school equivalency program

The services authorized by this subpart for the high school equivalency program include--

(1) recruitment services to reach persons--

(A)(i) who are 16 years of age and over; or

(ii) who are beyond the age of compulsory school attendance in the State in which such persons reside and are not enrolled in school;

(B)(i) who themselves, or whose parents, have spent a minimum of 75 days during the past 24 months in migrant and seasonal farmwork; or

(ii) who are eligible to participate, or have participated within the preceding 2 years, in [Migrant Education programs or JTPA §402 programs]; and

(C) who lack a high school diploma or its equivalent.

...

(c) Services provided by college assistance migrant program

(1)Services authorized by this subpart for the college assistance migrant program include--

(A) outreach and recruitment services to reach persons who themselves or whose parents have spent a minimum of 75 days during the past 24 months in migrant and seasonal farmwork or who have participated or are eligible to participate, in [Migrant Education programs or JTPA §402 programs], and who meet the minimum qualifications for attendance at a college or university;

...

34 C.F.R. §206.5(c) (1992):

(2) "Agricultural activity" means:

(i) Any activity directly related to the production of crops, dairy products, poultry, or livestock;

(ii) Any activity directly related to the cultivation or harvesting of trees; or

(iii) Any activity directly related to fish farms.

(3) "Farmwork" means any agricultural activity, performed for either wages or personal subsistence, on a farm, ranch, or similar establishment.

...

(6) "Migrant farmworker" means a seasonal farmworker--as defined in paragraph (c)(7) of this section--whose employment required travel that precluded the farmworker from returning to his or her domicile (permanent place of residence) within the same day.

(7) "Seasonal farmworker" means a person who, within the past 24 months, was employed for at least 75 days in farmwork, and whose primary employment was in farmwork on a temporary or seasonal basis (that is, not a constant year-round activity).

instructional services such as counseling and tutoring, housing support, assistance in obtaining financial aid, health services, exposure to cultural events, and a variety of other services.[55] As of November 1990, there were six CAMP grantees, located in five states.

The traditional definition governing eligibility in both HEP and CAMP differs significantly from the basic ME definition. Both migrant and seasonal farm workers (and dependents) are covered, but the worker must have spent a minimum of 75 days over the past 24 months doing farm work.[56] The regulations add a requirement that the worker's *primary* employment be farm work, and the farm work must be "on a temporary or seasonal basis (that is, not a constant year-round activity)."[57] Amendments in 1992, however, sought to improve cooperation between these two programs and other similar migrant programs, by expanding HEP/CAMP eligiblity to include persons currently eligible to participate in, or who have participated in either Migrant Education or the JTPA 402 program. For HEP that past participation must have occurred within the preceding two years.[58]

Migrant Health

Migrant Health is the oldest of the major federally funded migrant service programs. A House committee report explained the need for special federal measures, pointing to studies that "continue to show high infant mortality rates, high communicable disease rates, low prenatal care rates, high premature birth rates, high accident rates, low immunization levels, serious need for dental care, low economic and educational levels, mobility and lack of resident status leading to geographic and eligibility isolation from medical facilities, plus cultural factors and language barriers contributing to the health problems of migrant and seasonal agricultural workers and their families."[59]

In the early 1960s the Senate Subcommittee on Migratory Labor was considering several bills dealing with topics such as child labor, housing for migrant workers, and crew leader registration. Senator Harrison Williams, who chaired the committee, was approached with the suggestion to substitute a federally-funded service such as health services to help migrant workers rather trying to overcome farmer opposition to stricter regulation of the farm labor market. This strategy was adopted, and in September 1962 Congress passed without dissent a simple and straightforward bill adding a new section 310 to the Public Health Service Act. The law authorized up to $3 million "for paying part of the cost of . . . family health service clinics for domestic agricultural migratory workers and their families," as well as other special projects.[60] Congress appropriated only $750,000 for the first fiscal year of operation, but the program soon showed significant results, and congressional support grew apace, to $7.2 million in FY 1967 and $14.0 million in 1970. The FY 1991 appropriation was $51.7 million, up from $48.5 million in 1990.[61]

Because of limited funding, early efforts targeted preventive health services such as immunizations, health education, and environmental safety programs. Many recipients of federal grants also modeled their clinics on a program in Fresno County, California, which had emphasized accessible locations and evening hours so as to reach the migrant farm worker population more effectively. The clinics then relied to a large extent on referrals to local cooperating physicians.[62] Increased appropriations eventually allowed the provision of a wider range of services and the construction and later expansion of a network of clinics dedicated to serving migrants. Under the current version of the migrant health authorization, now located in section 329 of the Public Health Service Act, grants may be used for both ambulatory care and hospital services, as well as a host of other measures, where appropriate, including dental services, extended care, rehabilitative services, and necessary transportation.[63]

Until 1970, only migrant farm workers and their dependents were eligible for services under the Migrant Health program. An amendment that year added seasonal farm workers to the eligible population, because they face many of the same health problems and may "live side by side in the same community." Moreover, "their status as seasonal workers and as migrant workers frequently shifts back and forth."[64] Congress recognized that this change expanded potential eligibility manyfold, but the legislative history emphasized that the focus still remained on migrants. Services could be provided to seasonal agricultural workers and their families in a project only if the Secretary found that providing such services would contribute to the improvement of the health conditions of migrants. The conference report underscored this limitation.[65]

In 1975, Congress rewrote the Migrant Health authorization and added a great many detailed requirements for the establishment and operation of migrant health centers, including, for the first time, statutory definitions of "migratory" and "seasonal" agricultural worker.[66] Current law essentially follows the 1975 pattern, although there have been frequent refinements since then. The Act and the regulations retain a strong emphasis on service to migratory as opposed to seasonal farm workers, through a set of defined funding priorities that are linked to the number of migrants in the clinic's "catchment area."[67]

Who Is Eligible?

The Migrant Health program defines as migratory or seasonal agricultural workers only those whose "principal employment" is in agriculture on a seasonal basis.[68] (These characteristics are usually determined in clinics by intake workers, who generally rely on the individual's own statement about the primacy of agricultural employment; most do not spend much time or effort in checking on an individual's primary employment, for example, by calling employers or asking for payroll stubs.) A migrant is a person who establishes "a temporary abode" in order to be employed in agriculture, and to be eligible for MH services, an individual must have been employed in such work within the last 24 months. This 24-month look-back period is considerably shorter than that for Migrant Education, but of course migrants who have settled out for a longer period may still be served as seasonals, provided they retain their principal employment in agriculture. Such people will not count, however, for purposes of establishing the migrant population used in calculating funding priorities. "Agriculture," for Migrant Health purposes, includes only crops, not livestock, and it embraces processing, packing and similar activities only if "performed by a farmer or on a farm incident to or in conjunction with" primary growing or harvesting activity.[69] The committee reports accompanying the 1975 amendments, which adopted these definitions, offer no explanation of the reasons for choosing a 24-month period or defining agriculture in this fashion.[70]

In addition to the funding for Migrant Health Centers (MHCs) under section 329 of the Public Health Service Act, Congress began in 1975 a separate program of federally funded Community Health Centers (CHCs) for "medically underserved" areas, under a new section 330 of the Act.[71] The range of services that can be funded is quite similar to that available under section 329, and most migrant health catchment areas also qualify as "medically underserved." As a result, many Migrant Health centers also apply for and receive funding under section 330 as well.[72] There have been periodic efforts, most recently in the early Reagan administration, to repackage federal health care initiatives as block grants that would give the states far more discretion in deciding how to use their federal health funding. Congress has strongly resisted including migrant health in such packages, however, fearing that states would not accord sufficient priority to migrant programs if such a change were made.[73]

The Migrant Health Program is a branch of the Division of Primary Care Services in the Bureau of Health Care Delivery and Assistance of the Department of Health and Human Services (HHS). Its central office

---

Box 2.3 Definitions: Migrant Health

42 U.S.C.A. §254b(a) (West Supp. 1993):

(2) The term "migratory agricultural worker" means an individual whose principal employment is in agriculture on a seasonal basis, who has been so employed within the last twenty-four months, and who establishes for the purposes of such employment a temporary abode.

(3) The term "seasonal agricultural worker" means an individual whose principal employment is in agriculture on a seasonal basis and who is not a migratory agricultural worker.

(4) The term "agriculture" means farming in all its branches, including--

(A) cultivation and tillage of the soil,

(B) the production, cultivation, growing, and harvesting of any commodity grown on, in, or as an adjunct to or part of a commodity grown in or on, the land, and

(C) any practice (including preparation and processing for market and delivery to storage or to market or to carriers for transportation to market) performed by a farmer or on a farm incident to or in conjunction with an activity described in subparagraph (B).

42 C.F.R. §56.102 (1992):

(b)(1) *Agriculture* means farming in all its branches, including--

(i) Cultivation and tillage of the soil;

(ii) The production, cultivation, growing, and harvesting of any commodity grown on, in, or as an adjunct to or part of a commodity grown in, or on, the land; and

(iii) Any practice (including preparation and processing for market and delivery to storage or to market or to market or to carriers for transportation to market) performed by a farmer or on a farm incident to or in conjunction with an activity described in subsection (ii).

...

(h) *Migratory agricultural worker* means an individual whose principal employment is in agriculture on a seasonal basis, who has been so employed within the last 24 months, and who establishes for the purpose of such employment a temporary place of abode;

...

(m) *Seasonal agricultural worker* means an individual whose principal employment is in agriculture on a seasonal basis and who is not a migratory agricultural worker.

consists of four or five professionals, plus support staff, who have broad oversight and policy responsibility. The receipt and approval of migrant health center grant applications, as well as detailed program monitoring, are decentralized to the 10 HHS regions. Staff there report to the Regional Health Commissioners, not to the Migrant Health Branch of HHS in Washington. One staff member in each region is designated a Migrant Regional Program Consultant, but none of these consultants is able to devote full time to Migrant Health.[74]

Most of the annual appropriation for Migrant Health goes to the migrant health centers. The 102 MHC grantees operated some 400 clinic sites in 43 states and Puerto Rico in 1990.[75] Centers must undergo a competitive application process at least every five years, and must gain approval of "continuation" applications annually.[76] Some attention is given to geographic shifts in the migrant population in this application process, primarily through reviewing the past year's productivity statistics for the centers. Since productivity is likely to decline if the migrant or seasonal population in an area is shrinking, centers in such areas receive lower scores on their applications. But the persons we interviewed acknowledged the limitations of this system to respond to major shifts in migrant health needs. Limited funding in recent years has generally precluded the opening of new centers in previously unserved areas even if the central office knows of new migrant activity. There is no systematic arrangement for keeping track of such shifts in agricultural employment, particularly in areas outside the reach of existing centers.

Such a gap does not necessarily mean that migrants in these other areas will be without subsidized medical services. In many areas, health facilities locally known as migrant health clinics (including some in areas of high MSFW concentration in California and elsewhere) are not technically part of the federal Migrant Health program. These clinics receive no section 329 funds, relying instead on state and private sources, or on other federal support (such as section 330 or Medicaid).

In any event, MHCs themselves almost always rely on substantial funding from other sources (in addition to community health center funding under section 330 of the Public Health Service Act, for which a high percentage of MHCs qualify). They turn, for example, to state and local government grant or contract programs, church or other private support, Medicaid, private insurance, and patient fees.[77] Indeed, this search for supplemental funds reflects the great emphasis that the American system of service provision, relying as heavily as it does on private nonprofit organizations, places on such quasi-entrepreneurial initiatives. Those organizations that are most creative in finding new funding sources and perhaps -- in that process -- branching out into new but related fields of activity, will be best situated to sustain and expand their operations.

In addition to the funding of the MHCs, the central office of Migrant Health selects for special funding certain projects with wider impact. These include, for example, the East Coast Migrant Health Project, which recruits multilingual health care staff and outreach workers to work on a temporary basis during the peak season in various centers along the east coast, and the National Migrant Resource Program (NMRP), based in Austin, Texas. NMRP houses a library of studies and articles relevant to migrant health that can be drawn upon by all MHCs. NMRP has also spearheaded several other useful initiatives, including the development of a Migrant Clinicians Network and, under that umbrella, the generation of migrant-specific protocols to assist doctors and nurses dealing with migrants.[78] NMRP has also worked on issues of coordination among migrant health programs in various states, and with migrant service programs of other agencies. For example, its staff worked for many years to incorporate more complete and useful health records into the MSRTS data system. This effort ultimately failed, however, owing to database, confidentiality, and access problems, and NMRP is now looking to other techniques for providing easily transferable and readily usable medical records. Beyond these purely MH initiatives, the central office has arranged with the Health Care Financing Administration for pilot funding of hospitalization programs, primarily for maternal and emergency care, at selected locations.[79]

Coordination

Migrant Health has especially good working relationships with Migrant Head Start, for there is a natural match, in many localities, between the two programs. MH can provide the necessary health screening and treatment for children entering Migrant Head Start, and Head Start can also help MH to contact the children's families and so begin providing health services to them. In 1984, MH and MHS entered into a three-year interagency agreement to coordinate at the national level and to foster coordination at the local level.[80] Although the agreement apparently has lapsed, efforts are underway to renew it, and in any event a majority of MHS grantees report formal or informal agreements with MH at the local level.[81] In some states this kind of symbiosis has also flourished between MH and Migrant Education. Most of this coordination results from efforts at the local level, however, because MH is simply not staffed, at either the national or regional levels, to provide extensive initiatives for interagency coordination.

Nevertheless, MH has been highly supportive of better interagency coordination. It took the lead in efforts in 1985 to establish an interagency coordination body, and MH personnel and associated organizations have been quite vocal in calling for more structured coordination at the national level. The National Advisory Council on Migrant Health, for example, has called for the creation of "an Interagency Migrant Commission which exists at not less than the *Cabinet* level."[82] Sonia M. Leon Reig, Associate Bureau Director in the HHS Bureau of Health Care Delivery and Assistance, and formerly director of Migrant Health, testified before the National Commission on Migrant Education in favor of a Commission or Consortium, to be established at the "highest possible level, such as the White House," to "conduct short-term applied research, to prioritize and strategize solutions to common problems and to mobilize resources."[83] She also advocated a uniform definition of migrant farm worker, and she urged concerted efforts to eliminate duplication of activities, including mandated transfer of funds from one program to another to focus the responsibility for providing specific services.

Migrant Head Start

Project Head Start, of which Migrant Head Start is a component part, began in 1965 under the general statutory authorities granted by Congress to the Office of Economic Opportunity.[84] In 1969 OEO delegated responsibility for Head Start to what was then the Department of Health, Education and Welfare (HEW). Congress enacted a more specific statutory authorization in 1974,[85] and then revised and reenacted the "Head Start Act" as part of the Omnibus Budget Reconciliation Act in 1981.[86] The program is now administered by the Head Start Bureau, which is located in the Administration for Children and Families, Department of Health and Human Services (HHS).

Head Start is a comprehensive, locally based preschool child development program, often described as emphasizing five main components: education, nutrition, health, parent involvement, and social services.[87] Ninety percent of participants must be from families with incomes below the federal poverty guidelines, and at least ten percent of the spaces are reserved for children with disabilities.[88]

Head Start has been popular with both Democrats and Republicans, and, in an effort to help poor children, the program has been expanding. President Clinton reflected the conventional wisdom that Head Start can make a major difference in the lives of children who participate. Clinton justified expanding Head Start despite the budget deficit in his 1993 State of the Union Message by asserting that "for every dollar we invest [in Head Start] today, we'll save three tomorrow."

In FY93, a total of 720,000 preschool children are expected to be served at a cost of $2.8 billion; Clinton proposed a Head Start budget of $11 to $12 billion by FY97, meaning that the Head Start budget would have increased almost 9 times from its FY90 level of $1.5 billion.

The average regular Head Start program serves preschool children for about 4 hours per day and 4 days per week at an annual cost of $3,900, or, for 800 hours per year (fifty 16 hour weeks), Head Start spends about $5 per hour per child. Most Head Start programs are operated by nonprofit and public bodies, including schools, that apply for federal grants; in 1990, there were 1,320 Head Start grantees serving 540,000 children in 31,000 classrooms.

Special programs for the children of migrant workers were begun in the early years of Head Start, and the Head Start Reauthorization Act of 1969 required that Head Start programs for migrant farm workers and Indians be administered from the national level. Legislation in 1974 directed the Secretary of HEW to "continue the administrative arrangement responsible for meeting the needs of migrant and Indian children and [to] assure that appropriate funding is provided to meet such needs."[89] The same equally vague directive appears in the 1981 legislation and remains in effect today.[90] The statute authorizing Head Start now includes a funding formula that reserves 13 percent of the total appropriation for a list of designated priorities heavily (but not exclusively) oriented toward migrant and Indian children. In FY 1990, the allocation for Migrant Head Start was $60.4 million. Services were provided to 23,469 children in 33 states by the 23 Migrant Head Start grantees. In FY 1991, $74 million was made available, and in FY 1993, $85 million was available for MHS, so that 28,000 children could be served.[91]

It is much more difficult to calculate per child and per hour costs for children served by MHS. Many of the children in MHS programs are enrolled only part of the year; there are only about 11,700 full-year equivalent slots for children in MHS centers, generating an annual cost of $7,600 to serve a child 12 months in a MHS center. However, this cost-per-full-time-equivalent-MHS-child cannot be compared to the annual cost for serving a child in a regular Head Start program because MHS children are often served for many hours each day, 9 or 10 in MHS versus the more usual 4 in regular Head Start.

Program Operations

MHS grantees typically operate programs at numerous sites, either themselves or through delegate agencies, and some function in several states. For example, one grantee, East Coast Migrant Head Start, operates centers through delegate agencies in 12 states. Grantees are monitored in detail on a three year cycle, to measure accomplishments in light of detailed performance standards set out in the regulations and in contract documents. Head Start officials have asserted that this form of oversight, with a direct federal-to-local relationship, provides for better assurance (in comparison with the Migrant Education system) that services are provided effectively to the target population.

A typical regular Head Start center provides half-day programs throughout the school year. Because of its unique constituency, Migrant Head Start must operate in different ways and with greater flexibility. Some centers operate on a full-day basis, with two shifts, so as to provide services during the full time that the parents are working in the fields, and some put greater stress on center-provided transportation, nutrition, and even laundry services. In general, regular Head Start serves children only from age three to the age of compulsory school attendance; Migrant Head Start is authorized to serve children from birth to school age (usually 0 to 5 years of age). Recently 35 percent of enrollment in MHS consisted of infants and toddlers.[92] MHS also must remain flexible to provide its services during the time when migrants are in the area; time periods and demand may change from month to month and year to year, owing to shifting weather and crop patterns. Some personnel even move from place to place during the year with the migrant population.

Demand for MHS's day-care and educational services far outstrips supply. Many MHS programs maintain waiting lists for children from qualified families. Most operate on a first-come, first-served basis, provided that the family meets the migrancy and low-income requirements.[93] This arrangement gives an advantage to those families that are knowledgeable about the program and arrive early to assure sign-up. The advantages can be enormous. One MHS program had a staff of 57 during the peak summer months, to provide a most impressive range of services to the 81 children it can serve. The 57 include two shifts of teachers and aides, as well as bus drivers, outreach workers, cooks, laundry staff, and supervisory personnel. The center provides clothing for the children while in the center, washes their own clothes while they are there, and then sends them home in their own freshly laundered outfits.

Families who do not arrive in the area early enough to place their children in the Migrant Head Start program are probably relegated to day-care on a much more modest scale, sometimes day-care for which the family must pay, or they must make do in some other way. This lopsided outcome should raise legitimate questions about the priorities of the particular MHS program described here; the agency might be well advised to furnish less extensive services, in order to serve more of the needy farm worker population. In any event, as this example illustrates, MHS grantees have considerable flexibility to use their funds for direct educational and child care programs, and also for transportation, clothing, health care, and a variety of other support services.

Who Is Eligible?

The statute contains no definition of "migrant" or "farm worker" -- perhaps not surprising in view of the vagueness of the statutory provisions for Migrant Head Start in general. For most of the life of the MHS program, the regulations likewise contained no such definitions. Regulations containing a formal definition of "migrant family" were finally adopted in October 1992, but MHS officials stated that they merely codified the standard that has been used for years as a matter of administrative practice in a new definition of "migrant family." It is perhaps the most limited of the assistance program definitions, since only young children of current migrants are eligible for services. The look-back period is one year; the family must have moved in connection with agricultural employment within the past 12 months.[94] Only the production and harvesting of tree and field crops count as agricultural labor, and family income must come "primarily from this activity" if the family is to qualify for MHS.[95]

There has been some discussion among MHS directors about expanding the program to reach seasonal agricultural workers, and MHS recently chartered a limited pilot program for this wider constituency. But as long as funding remains limited, we found stronger support within MHS than elsewhere for retaining a relatively narrow definition focused on current migrants.[96] Some hold to this view even if definitions are harmonized across agencies, again as a way of targeting limited resources in a time of budgetary stringency. But some who advocated this view were also willing to countenance expansion to, e.g., a 24-month look-back period, because of the difficulties a family may face in its first year of settling out of the migrant stream.[97]

MHS performance standards require grantees and delegate agencies to coordinate with other available services at the local level.[98] As indicated above, cooperation is particularly in evidence with Migrant Health and other health agencies, in light of the specific obligation of MHS to provide health screening and certain health services to its children.[99]

Box 2.4. Definitions: Migrant Head Start

(42 U.S.C.A. et seq. §9831 (West 1983 & Supp. 1993))

[No statutory definitions.

57 Fed. Reg. 46,718, 46,725 (1992) (to be codified at 45 C.F.R. §1305.2(*l*)):

Migrant family means, for purposes of Head Start eligibility, a family with children under the age of compulsory school attendance who change their residence by moving from one geographic location to another, either intrastate or interstate within the past twelve months, for the purpose of engaging in agricultural work that involves the production and harvesting of tree and field crops and whose family income comes primarily from this activity.

Job Training for Migrant and Seasonal Farm Workers (JTPA 402)

Like Migrant Head Start, job training programs for migrant and seasonal farm workers originated in the Office of Economic Opportunity, under the very general language of Title IIIB of the Economic Opportunity Act of 1964.[100] In July 1973, however, responsibilities for MSFW job training and placement were transferred to the Department of Labor under a presidential order (supplemented by a Memorandum of Understanding between OEO and DOL), as part of the Nixon administration's efforts to phase out OEO.[101] Less than six months later, Congress completed work on long-debated umbrella legislation for federal job training and employment programs, the Comprehensive Employment and Training Act (CETA).[102] Section 303 of CETA, which became the framework for DOL efforts to assist MSFWs from 1974 through 1982, provided specific statutory authority for the ongoing programs to meet the training needs of migrant and seasonal farm workers.

CETA drew increasing criticism through the 1970s, primarily because of the operation of its non-migrant programs. Critics focused on CETA's extensive reliance on subsidized public-sector jobs, while at the same time audits disclosed high administrative costs and other operating problems. CETA thus became a prime target for the incoming Reagan administration, and the Act was allowed to expire in 1982. In its place Congress enacted a new framework for federal job-training programs, the Job Training Partnership Act (JTPA).[103]

As the title suggests, in its primary programs, the JTPA establishes a partnership between the private and public sectors covering all aspects of local policy-making and administration, including deciding locally what types and combinations of services to provide.[104] The general provisions of JTPA (separate from the farm worker provisions) give state governors several functions formerly assumed by DOL. In particular, the governors have a larger role in coordinating job training programs, and they designate local service delivery areas, the units of government within which the job training programs are to operate. Local programs operate under the guidance of local governments and Private Industry Councils (PICs), composed primarily of business representatives, but including members from labor, educational, and community groups. Most trainees are to be drawn from the ranks of the economically disadvantaged. Local programs must satisfy demanding performance standards, emphasizing successful job placements in unsubsidized employment. In addition, the Act places strict limits on the percentage of the funding that grantees may spend for administration.[105]

Despite persistent efforts from some quarters to bring MSFW programs under this decentralized JTPA umbrella, Congress chose to retain a special national program for farm workers, much like CETA 303, under § 402 of the JTPA.[106] Overseen by the Office of Special Targeted Programs in DOL,[107] rather than by PICs or state governors, JTPA 402 is not subject to the same partnership approach that characterizes general JTPA programs. Section 402 begins with a congressional finding that:

> "chronic seasonal unemployment and underemployment
> in the agricultural industry, aggravated by continual
> advancements in technology and mechanization
> resulting in displacement, constitute a substantial
> portion of the Nation's rural unemployment problem
> and substantially affect the national economy."[108]

The statute then authorizes services through public agencies and private nonprofit organizations that can administer "diversified employability development program[s]" for MSFWs.[109] Section 402 reserves an amount equal to 3.2 percent of the funding for Title IIA of the JTPA (the major general job training component) for MSFW projects, but Congress in recent years has appropriated funds above this level.[110] In addition, JTPA grantees have received funds as the conduit for other short-term federal assistance, such as money from the Federal Emergency Management Agency in response to a nationwide drought in 1988 and a California freeze in 1990-91. Congress appropriated $70.3 million for JTPA 402 in FY 1991, out of a total of $4.09 billion for all of JTPA.[111]

Program Administration and Operation

Following a procedure developed under CETA, but since refined, the distribution of JTPA 402 funds incorporates two steps. First, funds are allocated among the states based on population estimates of the number of farm workers in each.[112] Then a competitive process is used to decide on the grantee who will provide the JTPA 402 services in that state. DOL prefers to deal with a single grantee in each state, but California, with the largest number of farm workers, currently has five grantees.

The state-by-state allocations have been the subject of controversy for many years.[113] Current allocations are based upon the 1980 decennial Census of Population (COP), which shows the number of persons working in agriculture as of the last week in March. These COP data were adjusted in the late 1980s using Immigration and Naturalization Service data to account for legalized farm workers under the special amnesties enacted in the Immigration Reform and Control Act of 1986 for certain undocumented aliens.[114] Updated figures using data from the 1990 census will soon replace these. The use of COP data to distribute MSFW funds has been criticized, primarily because the census identifies mostly farm workers who were employed in March, the time of the census survey. Farm work is then at a low ebb; consequently this procedure underestimates the number of farm workers by at least two-thirds,[115] and it probably distorts the count in favor of home-base states over those where migrants may do most of their work during the growing and harvesting season.

Difficulties with decennial census figures reflect a larger problem with farm worker population data, as we will discuss in Chapter 5. Nevertheless, JTPA 402 grantees expect that COP data are likely to remain as the basis for their allocations. They have been working for years, so far unsuccessfully, to urge modest modifications in the COP questionnaires to better identify persons who have worked in agriculture. In particular, they have advocated a change in the census long-form questionnaire to ask that sample of respondents to distinguish between farm and nonfarm wages earned during the previous year.[116]

The regulations allow DOL to exclude states with small MSFW populations from Section 402 allocations if their total grant would be less than $120,000; for this reason, Alaska, Rhode Island, and the District of Columbia have no JTPA 402 program.[117] Other states qualify but are not guaranteed funding; potential grantees must compete on a biennial basis to be awarded the allocated funding to serve the state at issue. But in fact all the rest of the states are currently served, by a total of 29 nonprofit organizations, one local agency (Kern County, California), three state agencies (Florida, Utah, and Wisconsin), and the

Commonwealth of Puerto Rico.[118]   As the arithmetic indicates, some nonprofit grantee organizations have successfully competed to become the responsible agency for several states at once. Telamon Corporation, for example, serves nine states in the Midwest and East Coast regions.

JTPA 402 is a highly flexible program. Most of its funding goes for the training of those MSFWs who are seeking a major occupational change, and most such persons are seeking to leave agriculture for a more stable job in the nonfarm economy. Obviously persons who have settled out of the migrant stream, or are in the process of doing so, are the most likely to enroll in such programs, rather than current migrants. This result is not out of keeping with the program's purpose; JTPA 402, like its predecessors, specifically includes seasonals among those eligible for its services.

The services that may be provided to such MSFWs in connection with their retraining are extensive, and may include recruitment, assessment, classroom instruction, on-the-job training, job placement, follow-up and counseling, and other forms of support.[119]   The regulations require that at least 50 percent of a grantee's funds be spent on training. In addition, a wide range of expenses may be charged to "training-related support services," when provided to someone enrolled in this component of JTPA 402 -- for example, child-care, health services, financial counseling, and a stipend equal to an hourly wage to the individual during the time of the training.[120]   A grantee may spend several thousand dollars on a participant who makes use of this most extensive version of JTPA 402 services.[121]

But JTPA 402 is also designed to enhance the skills of those who choose to remain in agriculture,[122] and the most flexible component of the 402 program is directed to this end. The regulations stipulate that these "nontraining-related supportive services," provided to persons who are not enrolled in the more extensive training, work experience, or tryout employment programs, may include (but are not limited to) "transportation, health care, temporary shelter, meals and other nutritional assistance, legal or paralegal assistance and emergency assistance."[123] Grantees may not, however, use more than 15 percent of their grants for these supportive services. In 1990, 26,500 MSFWs received these supportive services, out of a total of 53,000 people served by JTPA 402 grantees.[124]

This element of JTPA programs is significant for interagency coordination.[125]   Other service programs may turn to JTPA when services that fall outside their own mandates are needed. For example, Migrant Health may ask JTPA for help in transporting an injured individual back to the home-base state after initial treatment at a clinic. But because eligibility for JTPA services is governed by a technical and restrictive definition of MSFW (discussed below), these requests occasionally cannot be honored, leaving the other service organization frustrated and probably impatient with definitional restrictions. Although such episodes are not terribly frequent and their impact should not be exaggerated, they accounted for the most common example cited to us during interviews of the ways in which definitions impose barriers to coordination.[126]

Definitions and Eligibility

Though §402 of the statute specifies "services to meet the employment and training needs of migrant and seasonal farm workers," it contains no definition of these key terms. Since 1974, definitions have been provided in the DOL regulations, frequently adjusted and refined as experience was gained. The 1974 version defined "farm worker" by reference to standard occupational classifications (SOC) provided in DOL's Dictionary of Occupational Titles.[127] This approach proved problematic, and in 1975 new regulations were issued, employing standard industrial classification (SIC) codes. These proved more workable, and all later regulations have maintained this basic framework.[128] Those who work for wages in agricultural production or in specified agricultural services are farm workers. Both crops and livestock are included, as is the on-farm packing of agricultural commodities.

Throughout the history of the regulations, migrant farm workers have been a subset of seasonal farm workers: migrants are those seasonals who are unable to return to their permanent places of residence within the same day. Thus, definitional refinements have focused on the criteria for "seasonal farm worker." The first such definition appeared in 1974, but was changed in 1975 "to assure consistency in the definitions used by different units of the Manpower Administration" of DOL.[129] At that time, the definition considered the person's employment only during the preceding 12 months, and individuals had to have worked at least 25 days in farm work but no more than 150 days in one establishment to qualify as a seasonal farm worker.[130] Grantees criticized this definition as overly cumbersome,[131] and then worked with DOL to produce a better one.

In 1979 another set of regulations was introduced, dropping the 150 day limit and replacing it with a simpler requirement that farm workers must not have a constant year-round salary if they are to qualify as seasonals. In addition, the look-back period was expanded to 24 months and an alternative to the 25-day minimum was provided. A farm worker would either have to work 25 days or earn at least $400 in farm work to qualify; this was seen as a "more realistic" criterion for the legislation's target population, and one that would be administratively feasible.[132] These specifications (but with a different look-back provision) survive in the current regulations.

In the rules proposed to implement JTPA § 402, after it replaced CETA § 303, DOL attempted to return to the 12-month look-back period.[133] This proposal drew considerable criticism, and DOL retreated to the complicated look-back compromise that appears in today's regulations.[134] In determining eligibility, grantees now must assure that the individual met the minimum farm work requirements during any period of 12 consecutive months during the past 24 months. They must also find that the individual was "primarily" employed in farm work.[135]

This is complicated enough. But the regulations also introduce other limitations, not as part of the definition, but as part of the eligibility requirements.[136] For example, during the eligibility determination period, the farm worker must have earned at least 50 percent of total earnings or been employed at least 50 percent of total work time in farm work. This appears to be an unnecessary limitation, since the definition already requires that a seasonal farm worker be "primarily" employed in farm work. Further, and more understandably, a means test is imposed. The farm worker must be part of a family either receiving public assistance or having an annual family income that does not exceed the higher of the poverty level or 70 percent of the Bureau of Labor Statistics' "lower living standard income level."[137] Dependents of farm workers who meet these stipulations are also eligible for the services of JTPA 402.

Finally, the regulations refer to general statutory limitations imposed by the JTPA on all recipients of services. These require that all males wishing to receive JTPA services must register with the Selective Service, and that all participants must be citizens, permanent resident aliens, or other aliens authorized to work in the United States.[138] These limitations sometimes prove frustrating for service providers in other programs trying to coordinate with JTPA 402, for their statutes do not

Box 2.5. Definitions: Job Training Partnership Act, §402
(29 U.S.C.A. §1672 (West 1985)) [No statutory definition.]
20 C.F.R. §633.104 (1992)):

Farmwork shall mean, for eligibility purposes, work performed for wages in agricultural production or agricultural services as defined in the most recent edition of the Standard Industrial Classification (SIC) Code definitions included in industries 01--Agricultural Production--Crops; 02--Agricultural Production--Livestock excluding 027--Animal Specialties; 07--Agricultural Services excluding 074--Veterinary Services, 0752--Animal Specialty Services, and 078--Landscape and Horticultural Services.

...

Migrant farmworker shall mean a seasonal farmworker who performs or has performed farmwork during the eligibility determination period (any consecutive 12-month period within the 24-month period preceding application for enrollment) which requires travel such that the worker is unable to return to his/her domicile (permanent place of residence) within the same day.

...

Seasonal farmworker shall mean a person who during the eligibility determination period (any consecutive 12-month period within the 24-month period preceding application for enrollment) was employed at least 25 days in farmwork or earned at least $400 in farmwork; and who has been primarily employed in farmwork on a seasonal basis, without a constant year round salary.

Id. § 633.107:

(a) Eligibility for participation in Section 402 programs is limited to those individuals who have, during any consecutive 12-month period within the 24-month period preceding their application for enrollment:

(1) Been a seasonal farmworker or migrant farmworker as defined in § 633.104; and,

(2) Received at least 50 percent of their total earned income or been employed at least 50 percent of their total work time in farmwork; and,

(3) Been identified as a member of a family which receives public assistance or whose annual family income does not exceed the higher of either the poverty level or 70 percent of the lower living standard income level.

(4) Dependents of the above individuals are also eligible.

impose similar limits, especially regarding legal immigration status. Moreover, some service providers, particularly those dealing with education,[139] feel strongly that they should do nothing to discourage undocumented alien farm workers or their families from taking advantage of their services. Nevertheless, JTPA grantees and DOL have no discretion in the matter, for these restrictions are express statutory requirements. Moreover, as several JTPA officials or service providers acknowledged, one cannot realistically expect a change, particularly in the requirement for work authorization, given that JTPA 402 is a training program designed to prepare workers for better jobs in the U.S. labor market.

Current Issues

During our interviews we heard some criticism of the JTPA 402 program, both from other service providers and from those involved in the program. Criticisms concerning coordination are mentioned above, as is the problem of using decennial census data to establish initial state allocations. We also learned that the strict performance standards employed by JTPA, with their emphasis on documented and successful job placements, can lead to "creaming." That is, grantees are induced to select for the training components of their programs only those already possessing the aptitude, talent or drive that might make them successful even without assistance, to the possible neglect of persons equally deserving but less likely to succeed. The same criticism is frequently directed to the general JTPA job training programs funded under the statute and run by PICs in the states and localities. Remedies are not obvious without relaxing some of the key accountability components on which Congress has insisted.[140]

Some persons we interviewed also commented that geographic targeting of services could be improved. Despite the apparent targeting provided by the two-step allocation process, many acknowledged that this only serves to provide a gross state-by-state population count. It does not necessarily assure that the state grantee will locate service centers in areas of heavy MSFW concentration nor keep up with changes in MSFW activity. Biennial grant reviews do monitor the effectiveness of services, and the federal contract officers may well pick a different grantee if the numbers provided by the first fall off. But if the grantee is serving well a significant number of persons at its current locations, Washington officials are not equipped to help identify other parts of the state that may not be well-served. One commented: "What do we know here about what's going on in the states?" More sensitive measures of

MSFW activities and populations would be needed to improve performance in this area.

Finally, some comments focused on the vigor or competitiveness of JTPA grantees. As indicated, most are private nonprofit organizations. Interviews indicated that they tended to be among the most entrepreneurial or aggressive of the grantee community. For example, several JTPA 402 grantees are also Migrant Head Start grantees, and a few also manage Migrant Health clinics. Many also pursue other private and public sources of funding, such as FEMA disaster assistance funds or community service block grant awards, and even AIDS education programs. Their success in these endeavors sometimes crowds out other assistance programs that apply for these funds and can therefore generate resentment. This may simply be an inevitable byproduct of the quasi-entrepreneurial competitive system used for choosing the providers of most of the MSFW services.

In any event, this enterprising approach may reflect the rather different nature of the task shouldered by JTPA 402, compared with those of other MSFW service programs. Although one should not over-generalize, a JTPA grantee may have more need to engage in affirmative outreach and persistent recruitment to enroll participants and thus meet its performance standards. Migrant Education has a natural location to find its target population, for migrant children are required to enroll in school. Migrant Health is an obvious destination when farm workers and family members become sick. Child care is also an obvious need when there are young children in the family, and Migrant Head Start easily becomes known as a provider. By comparison, JTPA's services are more remote and of uncertain application; MSFWs must often be persuaded to avail themselves of training opportunities.

Other Federal Programs Serving Farm Workers

The programs described so far (the "Big Four") meet certain important needs of farm worker families -- primarily medical care, education (including job training), and child care. Comprehensive assistance to such a family will often have to attend to other basic needs as well, such as food, housing, or legal assistance. Federal programs exist in each of these fields. Some specifically target the needs of MSFWs, but most of them are more general efforts to assist the disadvantaged. In the latter programs, farm workers who qualify can also take part, and on occasion such a program will adjust certain of its requirements or practices to take account of special requirements of migrants.

For example, nutrition assistance can be obtained through Food Stamps or the Supplemental Food Program for Women, Infants, and Children (WIC). The Food Stamp program, which dates to 1964 (and earlier pilot programs), is designed to increase the food purchasing power of persons with incomes below the poverty level and of those who are receiving certain forms of public assistance. Complicated formulas determine the amount of food stamps to which a household is entitled, and the stamps may be used to purchase any food for human consumption (alcohol, tobacco and imported foods are excluded). The program is overseen by the Department of Agriculture (USDA), but it is administered by state welfare or social services agencies.[141] In FY 1991 the program provided total benefits to recipients in the amount of $17.4 billion.[142] There is no special set-aside of funding for MSFWs, but in a realistic acknowledgment of the effect that uneven work patterns can have on farm worker families, the regulations make special provision for "expedited service" (ordinarily meaning receipt of food stamps within five days) to MSFW households in specified circumstances.[143]

The WIC program was adopted in 1972, primarily to address the issue of low birthweight babies. Administered by the Food and Nutrition Service of USDA, the program is implemented through state health departments. It now provides supplemental foods for a specified time period to pregnant, postpartum, or breast-feeding women and to children under age five, as well as nutrition education and certain health-related services. Recipients must meet income guidelines or else qualify for "adjunct eligibility" through receipt of food stamps, Medicaid, or Aid to Families with Dependent Children (AFDC), and they must be determined to be at "nutritional risk."[144] Recently Congress has insisted that the State plans under which WIC is provided must specially describe how they will address the needs of "migrants, homeless individuals, and Indians," and the Secretary is to report biennially on efforts to assure migrant participation despite their interstate movement.[145] The implementing regulations provide for a "migrant set-aside" of 0.9 percent of the fiscal year food appropriation for migrants, a total of $16.2 million in FY 1991.[146] WIC has also pioneered the use of a "verification of certification" card that is issued to migrant recipients to ensure continuity of benefits as the family moves from area to area. Those who have the card need not go through the whole process of application and eligibility determination in the new location. Service providers in other programs have expressed interest in adapting this model for other services, particularly with regard to Medicaid.

Housing needs are often acute for MSFWs. Over the past two decades, many employers have discontinued the provision of housing, in part because of greater success of enforcement of housing codes and other protective provisions administered by the Department of Labor.[147]

Substandard housing remains a major problem, and the need for improved enforcement continues. The federal government also provides funds to support the construction or rehabilitation of farm worker housing, through programs administered by the Department of Agriculture. Under § 514 of the Housing Act of 1949, USDA provides loans on highly favorable terms to farmers, farmers' associations, states, and private nonprofit agencies to construct or rehabilitate housing for farm labor.[148] Section 516 of the Act authorizes grants to nonprofit agencies to cover up to 90 percent of the cost of housing for the same basic purposes.[149] Funding for these purposes declined from $68.7 million in 1979 to $22.0 million in 1990, before rebounding to an appropriation of $27.3 million in FY 1992.[150]

The federal government provides civil legal services to persons who meet certain income and other criteria through a governmentally chartered private nonprofit body, the Legal Services Corporation (LSC), established in 1974.[151] A 1977 study, requested by Congress, established a special need for legal assistance to migrant farm workers, particularly in view of their usual remoteness from population centers, language difficulties, and frequent travel.[152] As a result, LSC undertook special efforts to create programs to meet these needs. In the mid-1980s, however, when this initiative appeared threatened, Congress provided a special line-item appropriation to assure continuation of migrant legal services. Migrant legal services programs now exist in 46 states, under an appropriation for FY 1992 of $10.8 million. Assistance using this federal funding may be provided only to farm workers with legal immigration status.[153]

None of these programs is able to serve what it considers its entire target population, and in any event each has limited capacity to assist those who are not physically located close to one of the program's facilities. When this is the case, the farm worker family must either do without the service, acquire it using family resources, or rely on state, local, or private funding to assist. At the state and local level there is enormous variety in the assistance of the latter types.[154]

NOTES

[1]See Philip L. Martin, Harvest of Confusion: Migrant Workers in U.S. Agriculture (1988).

[2]Generally speaking, seasonal farm workers are hired farm workers employed in agriculture on less than a year-round basis. Migrants are usually considered a subset of seasonal farm workers.

Some federal service programs focus principally on migrants; they are usually premised on the need for services to overcome the disruptions that derive from frequent changes of residence. But seasonal farm workers are also largely poor, and their intermittent employment also causes other sorts of disruptions and resultant disadvantages.

[3]According to the Department of Labor, Findings from the National Agricultural Workers Survey (NAWS) 1990, at 54 (1991) [hereinafter NAWS Findings], the median annual income for surveyed Seasonal Agricultural Services or crop workers was between $5000 and $7500 in 1990. The NAWs describes about 2.25 of the 2.5 million farm workers employed sometime during the year in U.S. agriculture. The NAWS defined migrants as workers who travel 75 miles or more to do farmwork, and because 40 percent of the NAWs sample workers spend part of each year abroad (usually in Mexico), id. at 83, at least 40 percent of the NAWs workers are migrants who shuttle back and forth across the U.S.-Mexican border. About 15 percent of the NAWs sample follow the crops in the United States, and some shuttle migrants also follow the crops, so that 42 percent of the NAWs sample are migrants, or 940,000 of 2,250,000 farm workers. If 940,000 workers earn an average $6,000 each, they earn a total $5.6 billion, roughly 10 times the amount now spent on federal MSFW service programs.

Another 40 percent of NAWs sample workers are employed less than year-round in U.S. agriculture. However, not all of these workers would be considered "seasonal" under all definitions: some do only a few days of farm work, and others do 9 or 10 months of farm work but do not work year-round. NAWS Findings, supra, at 77.

[4]See, e.g., House Comm. on Education and Labor, Elementary and Secondary Education Amendments of 1966, H.R. Rep. No. 1814, 89th Cong., 2d Sess. 10 (1966).

[5]See, e.g., Noel H. Klores, Farmworker Programs under the Comprehensive Employment and Training Act -- A Legislative History 24 (1981) (describing the reasons that led the directors of some of the earliest federal migrant assistance programs, under Title IIIB of the Economic Opportunity Act of 1964, to assure that migrant programs would be under centralized national management in the headquarters of the Office of Economic Opportunity (OEO)).

[6]Presidential Comm'n on Migratory Labor, Migratory Labor in American Agriculture 1 (1951).

[7]Pub. L. No. 89-10, 79 Stat. 27 (1965).

[8]Funding is based on figures showing low-income families, because of studies showing a high correlation between poverty and educational disadvantage. Once the funding is set, however, children are to be served on the basis of educational disadvantage alone, without regard to income.

[9]Pub. L. No. 97-35, tit. V, 95 Stat. 357, 463-82 (1981), replaced by Pub. L. No. 100-297, § 1003(a), 102 Stat. at 293 (1988) (current version at 20 U.S.C.A. §§ 2701-2976 (West 1990 & Supp. 1992)).

[10]Dep't of Education Fact Sheet, Allocations for School Year 1991-92.

[11]Elementary and Secondary Education Amendments of 1966, Pub. L. No. 89-750, 80 Stat. 1191. See generally Interstate Migrant Education Council (IMEC), Migrant Education: A Consolidated View 14 (1987); Congressional Research Service, Federal Assistance for Elementary and Secondary Education: Background Information on Selected Programs Likely to be Considered for Reauthorization by the 100th Congress 55-79 (prepared for the Subcomm. on Elementary, Secondary, and Vocational Education of the House Comm. on Education and Labor, Comm. Print, Serial No. 100-A, February 1987). Before 1966, a few initiatives for migrant education had been started under Title IIIB of the Economic Opportunity Act of 1964, Pub. L. No. 88-452, tit. IIIB, 78 Stat. 508, 525. See Sar A. Levitan, The Great Society's Poor Law: A New Approach to Poverty 249-61 (1969) [hereinafter Great Society].

[12]House Comm. on Education and Labor, School Improvement Act of 1987, H.R. Rep. No. 95, 100th Cong., 1st Sess. 36 (1987).

[13]The state directors of the 49 states where ME is active (all but Hawaii), have also established their own organization, the National Association of State Directors of Migrant Education (NASDME), to share information and ideas and to advocate better support for their programs.

[14]20 U.S.C.A. § 2783 (West 1990 & Supp. 1992).

[15]See 34 C.F.R. Part 205 (1992). See generally Congressional Research Service, Background Information, supra note 11, at 73-79.

[16]These were meant to correspond roughly to the western, central, and eastern migration streams. Several ME officials acknowledged that the streams are not so neatly divided any more, but they still believed that three regional offices made these services more accessible than would be the case with only one centralized coordination office.

[17]IMEC, Annual Report 1990-91, at 5.

[18]California, Texas and Florida apparently contribute additional funds.

[19]Once a child qualifies for ME, he or she can continue to be served in a ME even if the child moves across school district lines without parents. However, youth who migrate to do farmwork but who have parents who never migrated are apparently excluded.

[20]Elementary and Secondary Education Amendments of 1967, Pub. L. No. 90-247, § 109, 81 Stat. 783, 787 (1968).

[21]The Senate Committee explained the change to make formerly migratory children eligible for services with these words:

Children who have been left with friends or relatives while the parents are migrating to areas where work is available, suffer from a cultural gap when enrolled in the local school system even after receiving services in their first year of residence in a community. They continue to encounter difficult language problems and are reluctant to attend school because their attire may be shabby. They experience difficulty in becoming involved in the regular school community. These children have problems in adjusting to the alien cultural and sociological climate of the school system. The committee's amendments to title I will make possible the continuity of effort needed for special migrant programs to dislodge these children from the migrant stream and integrate them successfully into the local educational system.

Sen. Comm. on Labor and Public Welfare, Elementary and Secondary Education Act Amendments of 1967, S. Rep. No. 726, 90th Cong., 1st Sess. 13 (1967).

[22]20 U.S.C.A. § 2782(b) (West 1990); 34 C.F.R. § 201.3(b) (1992).

[23]Id. Another expansion of the statute's coverage occurred in 1974, when the definition was extended to include the children of migratory fishermen. Pub. L. No. 93-380, sec. 101(a)(2)(E), § 122(a)(1)(A), 88 Stat. 484, 492 (1974). This change was triggered by a highly successful migrant education program in Mobile, Alabama, in the early 1970s. That program turned out, upon inquiry, to be serving mostly the children of migratory fishermen, apparently on the misapprehension that such fishing was a form of agriculture. One of the key teachers, however, was the sister of a congressman, who was asked to introduce the new definitional language in order to allow the continuation of the Mobile program. Interview with Patrick Hogan, Office of Migrant Education, U.S. Department of Education, August 1, 1991. The conference report, however, explained the addition of fisheries to the definition as the

correction of a "technical deficiency." S. Conf. Rep. No. 1026, 93d Cong., 2d Sess. 142 (1974). Children of migratory fishermen now make up less than 3 percent of the population served.

[24]After the Reagan Administration proposed more restrictive definitions of some key terms, Congress mandated by statute the continuing use of earlier regulatory definitions. The current such provision appears in 20 U.S.C.A. § 2782(c) (West 1990).

[25]At the time of enactment in 1966, the Office of Education expected to be able to get such census estimates from health officials or the Office of Economic Opportunity. Allocation of all Title I education money was delayed while the figures were sought, but the delays stretched beyond what anyone anticipated. Finally a reserve figure was chosen in advance of census estimates, so that the balance of the Title I appropriation could be released to states and LEAs. Only a few weeks later did the office decide to use DOL statistics as the best available (although they admittedly covered only workers under contract, not all migrant farmworkers). The multiplier of .75 used to estimate the number of migrant children then derived not from scientific analysis but from the fact that its application to DOL statistics happily resulted in a total spending figure just below the amount of funds that had already been set aside by the Commissioner for Migrant Education from the overall Title I appropriation. Interview with Patrick Hogan, Office of Migrant Education, U.S. Department of Education, August 1, 1991.

[26]20 U.S.C.A. § 2781(b) (West 1990). The statute does permit the Secretary to use another system if he determines that it "most accurately and fully reflects the actual number of migrant students." The appearance of a rival to MSRTS under these conditions is thus highly unlikely.

[27]See National Comm'n on Migrant Education, Keeping Up with Our Nation's Migrant Students: A Report on the Migrant Student Record Transfer System (MSRTS) 2, 6-8 (Sept. 1991).

[28]See id.

[29]20 U.S.C.A. § 2781 (West 1990).

[30]Figures provided by the Office of Migrant Education. In very rough terms, the grants actually provided in program year 1991-92 worked out to a national average of about $500 per *identified* migrant student. This figure should be used, however, only to give an idea of the general order of magnitude of the funding received by states. Plainly the formula contemplates considerable variation state-by-state, and in any event, the funding is based on FTE calculations of identified migrant children in the state, not of those served.

[31]Office of Migrant Education Fact Sheet FY 1990, at 1 ("The number of migratory children served has grown from 80,000 in 1967 to 250,000 in 1981 and has remained fairly constant since then.").

[32]Maria V. Colon & Marlene Portuondo, Secondary Analysis of Selected Data on Migrant Education Programs, Fiscal Year 1990, Table G (Report prepared for the National Commission on Migrant Education, March 1, 1991). This table gives a figure of 359,996 served in the regular school year, as well as a figure of 126,796 served in summer school. Both figures contain an undetermined amount of double-counting, however, owing to the fact that some students receive services in more than one district over the course of the year.

[33]Testimony of Geraldine O'Brien, Executive Director, East Coast Migrant Head Start Project, before the National Comm'n on Migrant Education, April 29, 1991, at 4.

[34]Basing funding on population counts has not, for nearly a decade, resulted in added overall funding for the Migrant Education program. Ronnie E. Glover, the President of NASDME (the National Association of State Directors of Migrant Education) observed in a letter commenting on a earlier draft of this study:

"In theory our funding may be based on number of children identified, but in reality, since 1981, our funding is based on a line item appropriation. To the best of my knowledge, this is the same basis on which all other programs receive funding.

Letter from Ronnie E. Glover to Jeffrey Lubbers, ACUS (Feb. 7, 1992). (Even though added identification of migrant students probably will not increase the national appropriation, a potential for some distortion remains, because a state still can improve its relative position vis-a-vis other states in the annual allocations.) The Glover letter also expressed doubt that a change to funding based on actual numbers served would make much difference in state allocations, owing to the current requirement to serve all migrants, with the priority going to those in greatest need. Finally, it also voiced concern that such a change might instead force a reduction in services to particular individuals, especially to those in greatest need, as states tried to stretch resources to serve higher absolute numbers.

[35]The rules are slightly different for summer school, which is subject to special funding arrangements and counting rules under 20 U.S.C.A. § 2781(b) (West 1990).

[36]34 C.F.R. §§ 201.30 - 201.32 (1992).

[37]34 C.F.R. § 201.20(a)(3) (1992) (adopted in 1989). It appears that, in recent years Arkansas has obtained a special grant to operate its Hope camp rather than to count and enroll the children who pass through.

[38]These examples are provided by Robert Suggs, a consultant to the NCME. The life insurance information is based on brochures entitled: Washington State Migrant Education Program Accident Insurance. FL-18-90, (9/90); Accident Insurance Oregon Migrant Education. 1991-92; and USOE letter: ref: OCEP, of 7 February, 1974 from R.L. Fairley to J.O. Maynes, State Department of Education, Phoenix AZ.

[39]See Glover letter, supra note 34 ("All the objective data we have, including information from MSRTS and a comprehensive needs assessment in California, indicates that the needs of formerlies are, on average, at least 90 to 95 percent as great as those of currentlies.").

[40]The President of NASDME responded to such suggestions:

> Of course we want these "settled-out" children to be "picked up" by Chapter 1 basic and other programs, but in reality the Chapter 1 programs are often not the solution. For children who do not know English well enough to benefit from a Chapter 1 reading lab, or whose principal problem is that they are 16 years old and in the ninth grade, Chapter 1 is not a viable alternative. Even for those children whose needs could be addressed by Chapter 1 there is no assurance of access to those services; Chapter 1 can serve only about half the children who are eligible anyway.

Glover letter, supra note 34. (Any Chapter 1 program that is failing certain students in these ways, however, might well be out of compliance with the governing statute; it is not clear why the remedy should be continued ME coverage rather than specific remedies within the context of Chapter 1.)

[41]Office of Migrant Education, Fact Sheet FY 1990 (Table Headed "Final Allocation Migrant Education, in Rank Order, Program Fiscal Year 1990, 1201 Funds," June 6, 1990).

[42]34 C.F.R. § 201.3(b) (1992).

[43]For example, the Fresno, California ME program serves 22,000 ME students, including 10,000 Asians in Fresno city schools. Interview with Andy Rodarte, Migrant Education, Fresno, Calif. (Feb. 4, 1992). Farm employers and farm worker advocates assert, however, that Asians have never been even 5 percent of the area's migrant farmworkers.

[44]For a discussion of earlier battles over this same terrain, launched by an early Reagan Administration proposal to reduce the period from 5 years to 2 years, see Robert Lyke, Proposed Changes in Federal Programs

for Migrant Education 22-26 (Congressional Research Service, March 18, 1983). The proposal was not accepted by Congress.

[45]H.R. Rep. No. 95, supra note 12, at 37.

[46]See, e.g., National Migrant Resource Program, Integration and Coordination of Services at Migrant Health Centers III-24 (Feb. 28, 1992) (describing ME and MHS cooperation in programs involving the Illinois Migrant Council).

[47]20 U.S.C.A. § 2783 (West 1990 & Supp. 1993). This section also authorizes funding for the MSRTS system and sets aside a minimum of $6 million for these various coordination contracts.

[48]20 U.S.C.A. § 2782(a)(2) (West 1990 & Supp. 1992) (statutory cross-references omitted).

[49]In addition to HEP and CAMP, the Office also oversees the Migrant Even Start Program, a relatively recent addition to the list of educational assistance programs for migrants. 20 U.S.C.A. §§ 2741-2749 (West 1990 & Supp. 1993). As part of a larger Even Start program launched in 1988, it provides funds, through project grants to SEAs, to meet the special educational needs of migratory children and their parents by integrating early childhood education and adult education into a unified program. As of March 1991 Migrant Even Start had four grantees (in New York, Louisiana, Oregon, and Washington), and it received an FY 1991 appropriation of $1.5 million, up from $726,000 in FY 1990. (The legislation reserves three percent of the total appropriation for the migrant portion of Even Start. 20 U.S.C.A. § 2743(a)(1)(A) (Supp. 1993).) Eligible migrants include parents eligible for participation in an adult basic education program and their currently migratory children ages 1-7 inclusive. Formerly migratory children can also be included if space is available. See Office of Migrant Education, Directory of Services: Federal Agencies and Non-Federal Organizations Providing Services to Migrant and Seasonal Farmworkers and Their Families 11 (March 1991) [hereinafter Directory]; 34 C.F.R. §212.50 (1992). The relevant definitions, id. § 201.3, are the same as those used by the regular Migrant Education program.

The Department of Education is also authorized to provide adult literacy grants to the states under 20 U.S.C. § 1201 (1988). Under certain conditions the statute specifically authorizes the reservation of up to $3 million a year from appropriated funds to be used for "national programs," which includes, under id. § 1213, programs "to meet the special needs of migrant farmworkers and immigrants." See 34 C.F.R. Part 475 (1992). The definitions define "migrant farmworker" to include only those who have moved within the past twelve months. 34 C.F.R. § 460.4 (1992). In FY 1990 and 1991, however, no funds were made available for migrant adult literacy programs. Directory, supra, at 14.

Brief mention should also be made of a special program to provide vocational rehabilitation services for handicapped MSFWs and their family members, authorized by the Rehabilitation Act, 29 U.S.C. § 777b. Slightly over $1 million was appropriated for this program in 1990. Directory, supra, at 13.

[50]See Great Society, supra note 11, at 249.

[51]Pub. L. No. 95-524, § 2, 92 Stat. 1909 (1978), amending Pub. L. No. 93-203, § 303, 87 Stat. 839 (1973). See Robert Lyke, The College Assistance Migrant Program and the Migrant High School Equivalency Program 2-3 (Congressional Research Service, June 27, 1986).

[52]20 U.S.C.A. § 1070d-2 (West 1990 & Supp. 1993).

[53]In program year 1985-86, federal expenditures per student enrolled in HEP projects ranged from $905 to $3,842, with an average of $2,190. For CAMP the range was $1,819 to $5,900, averaging $3,038 per student. Lyke, supra note 51, at 5.

[54]20 U.S.C.A. § 1070d-2(b) (West 1990 & Supp. 1993); 34 C.F.R. Part 206 (1991).

[55]20 U.S.C.A. § 1070d-2(c) (West 1990 & Supp. 1993); 34 C.F.R. Part 206 (1992).

[56]See 20 U.S.C.A. § 1070d-2(b)(1)(B)(i). There has always been a significant group of young farm workers. For example, the 1983 Current Population Survey found that 35 percent of all farm workers were 14 to 17 years of age, and the NAWS found that about 17 percent of all farm workers were 21 or younger (perhaps 340,000 total). These young farm workers in many cases are not served by MEP, HEP, or CAMP. But at least HEP and CAMP definitions include youthful migrant workers themselves, whereas MEP generally covers only those whose parents or guardians move to seek farm work.

[57]34 C.F.R. § 206.5(c)(7) (1992). Migrants are defined as seasonal farmworkers "whose employment required travel that precluded the farmworker from returning to his or her domicile (permanent place of residence) within the same day." Id. § 206.5(c)(6).

[58]Pub. L. No. 102-325, tit. IV, § 405, 106 Stat. 507 (1992).

[59]H.R. Rep. No. 711, 91st Cong., 1st Sess. 2 (1969).

[60]Pub. L. No. 87-692, § 310, 76 Stat. 592 (1962). The background is recounted in Helen L. Johnston, Health for the Nation's Harvesters: A History of the Migrant Health Program in its Economic and Social Setting 135-39 (1985).

[61]National Association of Community Health Centers, Inc., Migrant Health Program Funding History: Fiscal Years 1963-1991 (mimeo 1991).

[62]See Johnston, supra note 60, at 151; General Accounting Office (GAO), Problems in the Structure and Management of the Migrant Health Program 3-4 (HRD-81-82, May 8, 1981).

[63]42 U.S.C.A. § 254b(a) (West 1991 & Supp. 1993).

[64]H.R. Rep. No. 711, supra note 59, at 3. The legislation that year was contained in Pub. L. No. 91-209, 84 Stat. 52 (1970). Before that time, migrant clinics had often provided services to seasonal farmworkers, but had to assure that funding for such services came from other sources.

[65]S. Conf. Rep. No. 853, 91st Cong., 2d Sess. 3 (1970), stated:
This provision is intended to be restricted in its applicability to projects in areas where migratory workers reside, and is to be limited to projects which will improve the health conditions of migratory workers themselves.

[66]Pub. L. No. 94-63, 89 Stat. 304 (1975).

[67]42 C.F.R. § 56.107 (1992). See also 42 U.S.C.A. § 254b(b) (West 1991 & Supp. 1993). Since 1978, migrant health clinics may also serve former migrants who no longer meet the definition because of age or disability. Id. § 254b(a)(1). Migrant Health published in 1990 an atlas of state profiles containing a detailed breakdown of migrant and seasonal farmworker population in each state served, as estimated (according to varying methodologies) by organizations in the states served by MH. Migrant Health Program, An Atlas of State Profiles Which Estimate Number of Migrant and Seasonal Farmworkers and Members of Their Families (March 1990).

[68]42 U.S.C.A. § 254b(a)(2), (3) (West 1991 & Supp. 1993).

[69]Id. § 254b(a)(4).

[70]See, e.g., S. Rep. No. 29, 94th Cong., 1st Sess. 106 (1975) (simply describing the definitions adopted in the legislation).

[71]42 U.S.C.A. § 254c (West 1991 & Supp. 1993).

[72]A GAO study reported that by 1979 63 percent of MH grantees were also funded under section 330. GAO, supra note 62, at 11.

[73]See, e.g., S. Rep. No. 618, 91st Cong., 1st Sess. 3 (1969). Johnston, supra note 60, at 167-68, quotes a Congressional statement announcing "unanimous agreement that the [federal migrant health] program had been successful, and that this success could be attributed to the program's separate identity that could be jeopardized by a merger with other programs."

[74]Interview with Jack Egan, Acting Director, Migrant Health Program (Oct. 21, 1991).

[75]Migrant Health Program (fact sheet, 1991); National Ass'n of Community Health Centers (NACHC), Medicaid and Migrant Farmworker Families: Analysis of Barriers and Recommendations for Change 1 (July 1991).

[76]Egan interview, supra note 74.

[77]Interview with Dan Cardenas, National Ass'n of Community Health Centers (October 21, 1991). No eligible MSFW may be turned away for inability to pay, but the centers may and do charge on a sliding scale that takes into account the resources of the patient or his or her family.

[78]See also National Migrant Resource Program, Migrant and Seasonal Farmworker Health Objectives for the Year 2000 (April 1990).

[79]Egan interview, supra note 74. This initiative reflects the difficulties migrants often have in making use of Medicaid, which should generally cover hospitalization costs for those who meet the low-income requirements, as migrants generally do. Migrants often find Medicaid difficult or impossible to employ, because of state residency requirements or simply because of paperwork delays that outlast their relatively brief stays. Reform of Medicaid to allow special eligibility requirements for migrant farmworkers, readily transferable as they move state to state, or to permit simplified application and approval, has therefore become a high priority for Migrant Health and affiliated organizations. See, e.g., National Advisory Council on Migrant Health (NACMH), 1990 Issues and Recommendations, at A-1 - A-2; Farmworker Health for the Year 2000: 1992 Recommendations of the National Advisory Council on Migrant Health 23; NACHC, supra note 75.

[80]Testimony of Frank Fuentes, Chief, Migrant Programs Branch, Administration of Children and Families, Dep't of Health and Human Services, before the National Comm'n on Migrant Education, April 29, 1991, at 17.

[81]Id. at 12, 17.

[82]NACMH, supra note 79, at A-4. In addition to NACMH, the National Council of Community Health Centers (NACHC) also serves as an umbrella organization watching out for the collective interests of migrant health clinics and attending to the need for better interagency coordination. In this respect these organizations are roughly the Migrant Health analogues of IMEC or NASDME for ME grantees.

[83]Testimony of Sonia M. Leon Reig, Dep't of Health and Human Services, before the National Comm'n on Migrant Education 10 (April 29, 1991).

[84]See Sar A. Levitan, Programs in Aid of the Poor 100-101 (5th ed. 1985); Economic Opportunity Act of 1964, Pub. L. No. 88-452, tit. IIIB, 78 Stat. 508, 525.

[85]Pub. L. No. 93-644, sec. 8, tit. V, 88 Stat. 2291, 2300 (1974).

[86]Pub. L. No. 97-35, §§ 635-657, 95 Stat. 499 (1981). The Head Start Act is codified at 42 U.S.C.A. § 9831-9855g (West 1983 & Supp. 1992).

[87]See 42 U.S.C.A. § 9833(a) (West 1983 & Supp. 1993); U.S. Dep't of HHS, Head Start: A Child Development Program 2-4 (1990). Under the statute and regulations, local policy councils, 51 percent of whose membership must consist of parents, exercise ultimate authority over personnel and fiscal matters. See 45 C.F.R. Part 1304, Subpart E (1993).

[88]45 C.F.R. §§ 1305.4, 1305.5 (1991).

[89]Pub. L. No. 93-644, sec. 8, tit. V, 88 Stat. at 2300.

[90]42 U.S.C.A. § 9831(b) (West 1983 & Supp. 1993). A 1990 amendment added "non-English language background" children to the list.

[91]42 U.S.C.A. § 9835(a)(2) (West 1983 & Supp. 1993). Telephone interview with Frank Fuentes, Chief, Migrant Programs Branch, Administration of Children and Families, Dep't of Health and Human Services (Sept. 30, 1991). An appropriation of $74 million for MHS in the FY91 budget of $1.95 billion indicates a 3.8 percent share for MHS. Data were updated in February 1993. Note that many migrant farm worker children spent only part of the year in a MHS program; there were about 2.4 children in each FTE slot (11,700 FTE slots enrolled 27,900 children). The average cost per FTE slot was $7,300 in 1993.

As with other farm worker assistance programs, MHS funds and enrollments are concentrated in a few states. California received about 20 percent of MHS funds in 1993, Texas 15 percent, and Florida 13 percent, for a 48 percent "Big 3" share. Washington received 9 percent of MHS funds, Oregon 7 percent, and Michigan 3 percent. All other states received 2 percent or less of total MHS funds (each 1 percent of MHS funding in 1993 was about $850,000).

[92]The 1992 regulations place a greater priority on selection criteria and careful choices about program priorities. 57 Fed. Reg. 46,718 (Oct. 9, 1992) (to be codified at 45 C.F.R. Part 1305. See id. at 46,721-22 (response to comments on proposed regulations, maintaining need for selection criteria in MHS program); 55 Fed. Reg. 29,970, 29,976 (notice of proposed rulemaking).

[93]Fuentes testimony, supra note 80, at 1-11.

[94]MHS considers a farmworker who has enrolled in a JTPA program to have "left agriculture," so that the child's eligibility for MHS services ends 12 months after the last qualifying move. JTPA 402 grantees would

generally like farmworkers who are enrolled in training programs to remain eligible for MHS services.

[95]55 Fed. Reg. 46,718, 46,725 (1992) to be codified at § 1305.2(l).

[96]Interview with Thomas Hill, MHS director, Fresno, California (July 16, 1991); interview with Geraldine O'Brien, Executive Director, East Coast Migrant Head Start Project (Oct. 22, 1991).

[97]Interview with Geraldine O'Brien, supra note 96.

[98]42 U.S.C.A. § 9837(c) (West 1983 & Supp. 1993) recently added a provision requiring all Head Start agencies to coordinate with local schools and other programs serving the relevant children and families.

[99]Through its National Migrant Headstart Directors' Association (NMHDA), an umbrella organization, MHS grantees have also cooperated in broader initiatives to improve interagency coordination.

[100]See Great Society, supra note 11, at 247-61; Klores, supra note 5, at 15-46. As amended in 1968, the statute described the program's purpose as follows: "to assist migrant and seasonal farm workers and their families to improve their living conditions and develop skills necessary for a productive and self-sufficient life in an increasingly complex and technological society." 42 U.S.C. §§ 2861-2862 (1970).

[101]See Klores, supra note 5, at 45-46; Ass'n of Farmworker Opportunity Programs, Toward an Equitable CETA 303 Allocation Formula for Farmworkers 4-5 (1978) [hereinafter AFOP].

[102]Pub. L. No. 93-203, 87 Stat. 839 (1973).

[103]Pub. L. No. 97-300, 96 Stat. 1322 (1982).

[104]See General Accounting Office, The Job Training Partnership Act: An Analysis of Support Cost Limits and Participant Characteristics 2 (GAO/HRD-86-16, Nov. 6, 1985).

[105]See New Job Training Program Replaces CETA, 38 Cong. Q. Almanac 40 (1982).

[106]Pub. L. No. 97-300, § 402, 96 Stat. at 1369 (codified at 29 U.S.C. § 1672 (1988)). The controversies over whether to retain a distinct, nationally-administered MSFW program or to incorporate it into the general, and decentralized, job training activities, recapitulated battles that have raged since OEO days. See, e.g., Klores, supra note 5, at 24-25, 48-66. Congress has generally supported national administration. Section 402(a)(2) now provides: "because of the special nature of farmworker employment and training problems, such programs shall be centrally administered at the national level."

[107]This Office also has responsibility for a few other national employment programs, such as those for Native Americans and older workers.

[108]29 U.S.C. § 1672(a)(1), 1988.

[109]Id. § 1672(c)(1).

[110]Department of Labor, The Farmworker Program (mimeo, Sept. 25, 1991). Amendments enacted in 1992 remove the 3.2 percent reserve provision, effective July 1, 1993 – thus, clearly leaving future funding levels to the appropriations process. Pub. L. No. 102-367, § 701(a), 106 Stat. 1103 (1992).

[111]AFOP Washington Newsline, July/Aug. 1991, at 3.

[112]20 C.F.R. § 633.105 (1992). Under this regulation, DOL first reserves 6 percent of the § 402 monies for a national account, usable for technical assistance and special discretionary projects. The balance, 94 percent, is then distributed among the states.

[113]See generally AFOP, supra note 101 (critiquing CETA allocation formula); California Human Development Corp. v. Brock, 762 F.2d 1044 (D.C.Cir. 1985) (court rejects extensive challenge to allocation formula).

[114]See Department of Labor, supra note 110, at 1; 55 Fed.Reg. 7,607 (1990).

[115]Leslie Whitener, Hired Farm Labor Data from the Decennial Census: Limitations and Considerations (Mimeo, August 1983).

[116]See testimony of Lynda Diane Mull, Association of Farmworker Opportunity Programs, before the National Commission on Migrant Education (April 29, 1991), at 4-5.

[117]Department of Labor, supra note 110, at 1.

[118]Id.

[119]Interview with Diana Carrillo, Center for Employment Training, Salinas, California (July 15, 1991).

[120]See 20 C.F.R. § 633.304(c)(3) (1992).

[121]The DOL reports that over the nine years of the JTPA 402 program, 97,000 farmworkers have been placed into permanent unsubsidized employment, at a unit cost of $3700. Department of Labor, supra note 110, at 2. It also reports a grand total of some 391,000 participants in employment and training activities generally, at an average expenditure of $1200 per participant. Id.

[122]Early versions of MSFW job training programs focused heavily on training for and placement in nonagricultural employment. This emphasis grew logically from the then-prevalent assumption that migrant farmwork was disappearing, increasingly displaced by mechanized harvesting. But it provoked the opposition of agricultural interests, who felt that these programs amounted to a federal effort to lure away a necessary workforce. Eventually this criticism was mollified by added statutory and regulatory language making clear that the job

training programs exist not only to train for nonagricultural employment, but also for enhanced employment within agriculture itself. See Donald B. Pederson & Dale C. Dahl, Agricultural Employment Law and Policy 136 (1981). For the current language to this effect, see 29 U.S.C. § 1672(c)(3) (1988).

[123]20 C.F.R. § 633.304(c)(4)(1992).

[124]Department of Labor, supra note 110, at 2.

[125]Through their principal umbrella organization, the Association of Farmworker Opportunity Programs (AFOP), JTPA 402 grantees have also paid considerable attention to interagency coordination issues. AFOP publishes a thorough and useful monthly newsletter, the *AFOP Washington Newsline*, and also a number of position papers and other studies.

[126]This frustration was probably magnified in the early 1980s, as job training programs shifted from CETA to JTPA. As indicated, JTPA imposed a 15 percent cap on these ancillary services; CETA had no equivalent limit. "With a regulatory limitation established on the amount of supportive services that could be provided by §402 programs, other agencies began to feel the burden of increased referrals because the programs could no longer support the cost of health, child care, transportation, and other emergency services at previous levels. The other major farmworker service providers were not expecting that such drastic changes would occur, nor were they expecting that their programs would now be required to pick up the pieces." Mull testimony, supra note 116, at 6.

[127]39 Fed. Reg. 28,401 (1974).

[128]40 Fed. Reg. 28,983 (1975). The current version, using virtually the same SIC codes, appears at 20 C.F.R. § 633.104 (1992).

[129]40 Fed. Reg. 28,980 (1975). Some of the key elements, including the 25 day minimum and 150 day maximum, track the definitions used for the ES-223 data reports elsewhere in DOL, as explained in Chapter 5. Persons interviewed noted this parallel, but did not believe that it had any real operative significance. In any event, later changes in the definitions for the job training programs have ended this commonality.

[130]Id. at 28,983.

[131]See, e.g., AFOP, supra note 101, at 66.

[132]44 Fed. Reg. 30,594 (1979).

[133]48 Fed. Reg. 33,182, 33,209 (1983).

[134]See 48 Fed. Reg. 48,774 (1983).

[135]20 C.F.R. § 633.104 (1992).

[136]20 C.F.R. § 633.107(a) (1992).

[137]The latest revision of the lower living standard income levels, provided solely for JTPA eligibility purposes, appears at 58 Fed. Reg. 15,506 (1993).

[138]20 C.F.R. § 633.107(c) (1992), referring to sections 167(a)(5) and 504 of the JTPA, 29 U.S.C. §§ 1577(a)(5), 1504 (1988).

[139]See generally Plyler v. Doe, 457 U.S. 202 (1982) (Constitution forbids certain state restrictions on education for undocumented alien children).

[140]See, e.g., Kirk Victor, Helping the Haves, 22 Nat'l J. 898 (1990); Robert Guskind, Cheers and Bronx Cheers for Jobs Law, 20 Nat'l J. 2407 (1988).

[141]See generally Levitan, supra note 84, at 77-81; David A. Super, Introduction to the Food Stamp Program, 23 Clearinghouse Rev. 870 (1989). The Food Stamp Act is codified at 7 U.S.C. §§ 2011-2030 (1988), and the implementing regulations appear at 7 C.F.R. Parts 271-282 (1993).

[142]Telephone interview with Daniel Woodhead, USDA Food and Nutrition Service (April 15, 1992).

[143]See 7 C.F.R. §§ 273.2(i)(1), 273.10(e)(3) (1993).

[144]See generally Staffs of the American Civil Liberties Union and the Food Research and Action Center, Introduction to the WIC and CSFP Programs, 24 Clearinghouse Rev. 820 (1990). The authorizing legislation is codified at 42 U.S.C. § 1786 (Supp. 1989), and the implementing regulations appear at 7 C.F.R. Part 246 (1992).

[145]42 U.S.C. § 1786(f)(1)(C)(4), (j) (1988 & Supp. 1989).

[146]Telephone interview with J.B. Passino, Food and Nutrition Service, USDA (April 16, 1992). See 7 C.F.R. § 246.16(c)(2)(i)(B) (1993). The regulations define "migrant farmworker" as "an individual whose principal employment is in agriculture on a seasonal basis, who has been so employed within the last 24 months, and who establishes, for the purposes of such employment, a temporary abode." Id. § 246.2.

[147]See General Accounting Office, Hired Farmworkers: Health and Well-Being at Risk 28 (GAO/HRD-92-46, Feb. 1992).

[148]42 U.S.C. § 1484 (1988).

[149]42 U.S.C. § 1486 (1988). The implementing regulations for both the loan and grant programs may be found at 7 C.F.R. Part 1944, Subpart D.

[150]GAO, Hired Farmworkers, supra note 147, at 29; Telephone interview with Tom Sanders, Multi-Family Housing Division, USDA (April 16, 1992).

[151]42 U.S.C. §§ 2996-2996*l* (1988).

152See General Accounting Office, Legal Services Corporation: Grantee Attorneys' Handling of Migrant Farmworker Disputes with Growers 2 (GAO/HRD-90-144, Sept. 1990).

153See 45 C.F.R. § 1626.4 (1992).

154The above listing does not exhaust the range of federal assistance programs that might possibly be called upon in meeting the comprehensive needs of a farmworker family. See generally Table 1, supra, and Directory, supra note 49. In addition to assistance programs, the federal government has established several enforcement regimes that can be brought to bear to improve the situation of MSFWs. The most important are the Migrant and Seasonal Agricultural Workers Protection Act, 29 U.S.C. §§ 1800-1872 (1988), the Fair Labor Standards Act, 29 U.S.C. §§ 201-219 (1988), and the Occupational Safety and Health Act, 29 U.S.C. §§ 651-678 (1988), all three administered by the Department of Labor, and the Federal Insecticide, Fungicide, and Rodenticide Act, 7 U.S.C. §§ 136-136y (1988), administered by the Environmental Protection Agency.

3

Coordination Among Migrant Assistance Programs

State Level Coordination

Most of the MSFW service providers and officials we interviewed felt that local and state level coordination in their areas had improved in recent years, but nearly all agreed that more could be done. Much of the recent improvement in coordination has stemmed from statewide task forces or councils on migrant farm workers, usually established under the authority of the state governor. Typically these bodies bring together a wide range of interests, ranging from growers' representatives through officials of service and enforcement programs (both MSFW-specific and more general programs) to Legal Services and farm worker advocacy groups.

In Virginia, for example, after a somewhat acrimonious start many years ago, the Governor's Board on Migrant and Seasonal Farmworkers moved beyond adversarial relationships to focus on cooperative initiatives, such as the construction of improved housing for migrants.[1] In Illinois, the statewide Inter-Agency Committee on Migrant Affairs meets six to eight times a year to set priorities, share analyses of farm labor trends, identify gaps in services, and undertake similar functions. Local networks are also encouraged to get together regularly during the peak migrant season.[2] In Indiana, a Task Force on Migrant Affairs has existed since 1952; it meets monthly and has numerous standing committees to examine specialized issues. One of these committees spearheaded the creation of an impressive Consolidated Outreach

Project to enroll migrants and their children in federally-funded assistance programs.

Some of these programs merit fuller description, because they could serve as models for similar efforts elsewhere. In Indiana, many MSFW programs faced declining budgets in the early 1980s. Seeking ways to do more with less, the Program Operations Committee of the Indiana Task Force proposed pooling agency funds used for intake processing and for outreach (e.g., to locate new concentrations of MSFWs or facilitate their access to services). Although the different service programs used varying definitions of who was eligible for services, it proved possible to develop a single-page form that would obtain the main information needed by each of them. Four agencies initially agreed to cooperate, and the Indiana Department of Human Services won the initial contract to provide these outreach services, but recently this function was shifted to Indiana Health Centers, Inc., a private nonprofit organization that is also the Migrant Health grantee for the state. That organization now has 23 caseworkers on its staff performing this outreach function throughout the state.

The consolidated outreach form has been refined over the years, so that the caseworker can go over each item carefully with the farm worker being interviewed, usually at his or her residence in the migrant camp. After the form is completed, both caseworker and interviewee sign it. One copy of the form is used to enter the data in a central computer system, which keeps a complete individual record and also generates limited monthly census data usable for funding purposes. Other copies go to various programs or caseworkers, and one copy is kept by the farm worker family. The Project uses financial incentives to encourage clients to keep the form and make use of it as they obtain services from MSFW programs. For example, certain discounts on Migrant Health services are available for those who present their own yellow copy at the clinic, and this copy also gains them preferred access to a migrant food pantry.

The consolidated outreach process has not fully achieved its original objectives; it does not entirely replace individual intake processing by the separate programs. The originators of the program eventually came to appreciate the additional functions, in addition to form completion, that the intake staff of the various programs performs; hence some intake staff for each of the specific programs had to be retained. For example, these staff members also carry out preliminary needs assessments and furnish counseling. Nevertheless, we were told that the consolidated outreach form saves an average of 45 minutes per case for the intake staff of specific programs -- a worthwhile economy. Consolidation also reduces considerably the burden on the MSFW family seeking to use several of the available services.[3]

Iowa has achieved coordination through a different process. Proteus, Inc., had been the JTPA 402 grantee for the state for several years when, in 1990, it also successfully competed to become the state grantee under the Migrant Health and Migrant Head Start programs. The management team thereupon decided to use a single intake staff for the basic intake and outreach process to enroll MSFWs or their dependents in any of the three programs. But Proteus caseworkers soon encountered a problem. Under the statute, JTPA processing requires detailed information about the immigration or citizenship status of an enrollee, whereas MHS and MH have insisted on agnosticism about such issues, in order better to fulfill their underlying missions. Eventually Proteus decided to have phased questioning, and to train its staff to begin all intake sessions with a brief orientation, during which prospective enrollees are counseled that if they desire only medical or Head Start services, they will not have to answer questions about immigration status. Persons not seeking JTPA services also are subject to less rigorous documentation requirements; for example, they need not necessarily show W-2 forms to document work history.[4]

The Proteus intake form is a one-page sheet that allows for the gathering of a considerable amount of background information, including work history, that can be used in determining eligibility for a number of programs, and also to help determine specific needs within the programs. In fact, Proteus also uses this form as the basis for preparing certificates of eligibility for the state Migrant Education program, a task it carries out under contract with the state ME office.[5]

Further consolidation or integration of actual services has been hampered, however, because the parent federal agencies insist on keeping their own component of Proteus' services separately identifiable, somewhat like a stand-alone program. The great variety in the performance standards of the programs also inhibits comprehensive planning, even if one nonprofit agency operates several assistance programs. For example, Proteus strives to avoid turning anyone away who seeks child care services. Nevertheless its MHS program, which operates during the peak farm work months of July and August, has a fixed enrollment limit. When needed, additional child care is provided, on a modest scale, using nontraining support services funds from JTPA. MHS in particular is said to be hard to administer because of its elaborate requirements for plans and reports, and its demanding performance standards. For example, MHS requires that newly-enrolled children have full physical and dental examinations, even if the child received such screening during the past month in another program. Some greater comparability among the programs in this respect would promote greater integration of services.

In Illinois, the JTPA 402 grantee, the Illinois Migrant Council (IMC), also became the Migrant Health grantee. In addition to the opportunities for consolidation and coordination which this arrangement afforded, further coordination has succeeded with a number of other state and private bodies that serve farm workers. According to a recent thorough case study of the program, IMC has developed cooperative working relationships with county health departments, the Department of Rehabilitative Services, Department of Public Aid, Migrant Education and Migrant Head Start, the state health department, the American Medical Students Association, the Rural Community Assistance Program, and the statewide Inter-Agency Committee on Migrant Affairs.[6]

Particularly impressive in Illinois are the formal agreements negotiated with ME and MHS to encourage highly integrated medical services to migrant children. Physical exams of school age children, for example, are performed by IMC, with ME paying approximately 50 percent of the cost. Dental services are also provided by IMC, at a fixed rate of $20 per child for MHS participants and under a flat-fee arrangement with ME for $15,000 to serve approximately 1200 school-age children.[7] This combined effort has also facilitated a comprehensive strategy to deal with dental problems among MSFWs. Arrangements have been made for sharing of records throughout the season, and a season-end "record swap" to assure complete documentation on all children.[8]

The Cornell Migrant Program based near Rochester, New York, spearheaded the development of a Working Together Group involving eight MSFW service programs in Western New York. It included the Big Four programs as well as legal services, a literacy program, and a social ministry. The Group was formed under the guidance of an outside facilitator in 1988 to reduce conflicts among agencies that serve MSFWs as well as to plan for a conference that year to address racism against minority migrant workers.

The planning process led to coordination. In 1989 the agencies successfully cooperated to stage a farm worker festival. Later they worked to develop a coordinated outreach effort involving a joint intake form and joint training for outreach staff. The MSFW agencies then cooperated to win a grant to deal with substance abuse among farm workers, and another to coordinate literacy services. The guiding principle of the Group appears to be coordination to obtain additional resources for joint activities, usually in areas that fall outside the reach of the specific mission of each program, rather than endeavors that might ultimately lead to transfer of funding from one agency to another as pre-existing tasks are consolidated.

Coordination at the Local Level

Often statewide task forces or councils mandate or encourage the creation of similar local service providers' councils. There are several in Virginia, for example; the council on the Eastern Shore, where MSFW activity is concentrated, meets monthly and is working on developing a consolidated outreach approach. Local councils also provide a forum for discussing common issues and sharing information about program or about changes in the population or farm work patterns.

In many areas, 5 to 10 outreach workers descend on migrant workers and their families as they settle into public or well-known private camps. Migrants truly new to the system are often bewildered; after often arduous journeys and settling into modest temporary quarters, they are offered full-day child-care at no charge, summer school for their children, and health and legal services. To minimize such confusion and duplication, many local coordination bodies have begun to establish "service fairs," in which all of the migrant programs visit a camp simultaneously, set up tables to provide information on the services each provides, and thus make it easier for migrant families to learn about and register for the programs in which they might be interested.

Coordination: Promises and Problems

The primary function of existing coordination forums is information sharing. They may also facilitate the process whereby the various programs refer a client to another program -- a sick child may be taken from an ME program to a migrant health clinic, for example. Coordination runs into problems, however, when the agencies must deal with issues that may have resource implications.[9] For example, local or state ME and MH agencies may argue, after a referral, over who bears responsibility for paying for the health services provided to school children. As a result, in some cases, ME programs employ their own medical staff to care for children enrolled in ME programs rather than using MH staff and facilities. Similarly, coordinated outreach proposals have run aground on these sorts of financial issues, such as when the agencies cannot agree on a formula to pay for the cost of the single primary outreach staff.

Sometimes resource conflicts become so acrimonious that they may inhibit recognition of productive coordination -- giving rise to a misleading sense that less ambitious coordination efforts, such as those that do not seek any resource transfers among programs, are more successful because they give rise to fewer complaints. For example, a recent national meeting of the major MSFW service providers awarded special recognition to the Working Together Group from New York, selecting it over the Indiana Coordinated Outreach Program. The New York program, relying on a Cornell University-funded staffer to serve as a neutral convener to help work out common problems and to avoid competition for funds, generated less opposition than the Indiana program. The nomination of the Indiana program, in particular, was resisted by the former JTPA 402 grantee in the state, which felt that its needs had not been adequately met by the combined system, resulting in reduced funding for that JTPA grantee and eventually the termination of its 402 grant.[10]

The disinclination of coordination bodies to deal with resource reallocation issues is understandable, but unless coordination is expanded to deal with shared funding, it will be difficult to see how coordination can lead to the efficient provision of comprehensive services. Although it is not easy to create coordinating bodies that can also deal with resource shifting, it seems clearly advantageous for those states currently lacking a state-level coordinating body to create one, with representation from all interested parties, both public and private. This coordinating body should have a specific mandate to examine resource issues, with authority to recommend changes in service allocations. It should also encourage well-focused local coordination efforts, perhaps by rewarding successful coordination efforts with additional funds.

Do Differences in Definitions Impede Coordination?

Many first-time observers of the social service scene are surprised that there are so many programs, each with their own outreach workers, intake forms, and facilities to provide services. One contributing cause for this seeming duplication of administrative overhead is the fact that each program has a different definition of who is eligible for services. Because program auditors sometimes require programs to return the money that they spent on ineligible clients, programs understandably want to guard against such audit findings with their own personnel.

One seemingly logical way to promote coordination among programs that provide services to migrant workers and their families would be to have a common definition of migrant farm worker. And when asked in the abstract about obstacles to coordination among MSFW assistance programs, service providers and officials commonly cite the varying definitions that govern in the separate regimes.[11] But our field interviews failed to turn up widespread evidence of significant concrete problems caused by the various definitions of migrant farm worker.

Many local service providers had to think long and hard before coming up with examples of how differences in definition impeded coordination, and many of the examples seemed to be nuisances rather than systemic barriers to coordination. For example, MH and MHS sometimes were frustrated that some of their clients could not make use of JTPA 402 vans or other transportation services, because of JTPA's more restrictive eligibility criteria. Others noted that different definitions make it harder to consolidate outreach and intake processing, and may discourage the programs from even attempting such cooperation. Several persons interviewed noted the wastefulness of sending numerous outreach workers to burden a farm worker family, since each asked many of the same questions, even if each worker does spend part of his or her time asking certain questions that are germane only to the particular assistance program at issue. The Indiana and Iowa efforts indicate, of course, that considerable progress can be made in consolidating intake processing, even while different definitions govern the various programs. But the psychological barrier remains.

Other consequences of definitional differences were also mentioned by some of the people that we interviewed, but it proved difficult to pin down specific details. We were told that cooperation between programs sometimes foundered because of "political fallout" once farm worker parents learned that some of the children would be excluded from one portion of a combined program, owing to different eligibility standards. For example, we heard of a Migrant Education program's agreement to facilitate the efforts of the local Migrant Head Start by offering space in a summer school building and permission to use ME buses for transportation. The Migrant Health clinic agreed to provide health screening and inoculations. But when the ME buses went out to pick up children for both the ME and the MHS programs, the driver could not be reimbursed from MHS for transportation if he also transported "formerlies." Angry parents called the school board, which then decided that the cooperative effort should be discontinued.

Obviously in this case definitional differences were not an absolute barrier to coordination; it should have been possible to sustain the combined program, given additional effort to explain the situation, ride out the immediate negative reaction, or provide alternative assistance of a similar type to those excluded from MHS. But the differences did complicate matters. And viewed in a larger perspective, it may not make sense to have two educational efforts for young children, often siblings from the same family, treating them as markedly different constituencies.

Based on such a line of argument, some people that we interviewed argued for a procedure whereby coordinated programs could overcome such problems by means of a waiver procedure. They proposed that by qualifying for one of the cooperating programs, an individual (or family) could have access to all the others, whenever local service providers in the various programs negotiated arrangements for coordinated or integrated services. In the above example, children could ride the ME buses to the MHS program if their families met the relevant definition for *either* ME or MHS. Or to pursue the example further, a seasonal farm worker enrolled in a lengthy training program under JTPA 402 would be eligible for child care at the local MHS center, even though it had been more than a year since he or she last migrated to undertake agricultural work. Granting such waivers would probably require statutory amendment. [12]

This cross-eligibility or waiver proposal holds some initial attraction, but it also gives rise to important questions. The net result would clearly be an expansion in the population eligible for any one of the given programs. Unless coupled with either a major funding increase or some other rationing mechanism to replace the rationing function of the original definitional limitations, it might only lengthen waiting lists, exacerbate uneven service, or dilute the level of assistance to the primary target population of a program. Moreover, it would amount, in practice, to a kind of uniform definition, but one that incorporates the most expansive features of each of the programs' definitions.

If there is to be some such de facto uniform definition, it would be better to have it result from direct decisions on each of the elements of the definition. This could mean that sometimes expansive criteria are chosen, such as in the lookback period, and sometimes narrower ones are selected as a way of better targeting limited resources, such as restricting services to workers who work on farms rather than in processing plants. Cross-eligibility without agreement on a definition might also encourage some manipulation or reward the canniest applicants for services. For example, a seasonal farm worker who had never migrated but who wanted to enroll his children in MHS or ME could achieve this objective by first qualifying for MH or JTPA benefits, both of which include seasonals. Other seasonal farm workers who applied directly to ME or MHS, however, would not be eligible, at least not without following the circuitous route of the first.

These problems should give serious pause before launching a waiver or cross-eligibility procedure without considering a uniform definition. Nevertheless, it is not clear just how substantial they might be in practice; some interviewees speculated that enough resources might be saved by the elimination of duplicative eligibility determinations to pay for the additional services that might be required for the additional clients such a procedure generates. It might therefore be worthwhile to test this waiver proposal in the field by a more limited statutory amendment, allowing the designation of some local areas for pilot projects. The task of selecting the pilot locations and working out the exact ground rules for such waivers could be assigned to a federal coordinating entity.

In any event, even complete uniformity in the federal definitions would not usher in a new era of simple interagency coordination, for a very straightforward and important reason. From the perspective of the service provider in the field, seeking to help clients to draw on other services available locally, these federal programs form only part of the picture. Most localities do not have all of the Big Four assistance programs. If there is no Migrant Health clinic locally, ME or MHS personnel seeking health care for one of their students will probably have to work with a local physician's organization, a local hospital, or perhaps with a state-funded health clinic, or they may turn to private sources of support. Many of these agencies or organizations will have their own eligibility criteria, which may or may not include a definition of MSFW. If a client does not qualify for one of these sources of assistance, effective local service providers do not spend a lot of time grumbling about definitions; they simply go on to look for another source of support.

Given the diversity of assistance programs available under our federal system, therefore, coordination will have to occur primarily at the local level, taking into account the full range of relevant programs available in that location, both MSFW-specific and general, and both public and private.[13] State and national coordination initiatives can still be useful, however, to promote such local initiatives and to find incremental ways to overcome existing barriers. Constant and creative prodding from such quarters can also help to stimulate local initiative and to overcome personality conflicts. In sum, as many of the persons we interviewed observed, the absence or ineffectiveness of local and state coordination probably has more to do, overall, with lack of local initiative or personality conflicts than with structural barriers; differences in definition can serve more as an excuse than an explanation for a lack of coordination.[14]

In order to promote coordination between state and local programs that provide services to migrant workers and their families, we recommend the creation or improvement of state-level coordinating bodies. These bodies should look for statewide initiatives that can make more efficient use of available resources. These bodies should also attend to ways to promote better local-level integration or cooperation. As to definitions, we cannot conclude that the current diversity in federal definitions imposes a truly significant barrier to coordination. Coordination is clearly possible without a uniform federal definition of who is to be served, and considerable local initiative for coordination would still be necessary, even if there were a uniform federal definition, given the diversity of local resources available.

Moreover, differing federal definitions took root, in part, for understandable reasons relating to the specific service missions of the varying programs. Immediate mandates to force definitional uniformity are therefore likely to provoke considerable political resistance. We were frequently reminded during interviews, for example, of a Reagan administration proposal to cut the Migrant Education look-back period from five to two years, thus limiting the number of "formerly migratory" children eligible for ME services. After acrimonious controversy, this more restrictive definition, which was more in line with other definitions, was successfully beaten back by the affected agencies, and Congress reaffirmed the five-year period.

Despite these cautionary notes, we nonetheless recommend steps in the direction of a uniform definition, for two reasons. First, we believe that consolidated outreach offers real hope for improved service to individuals, both by cutting down on the wasteful use of staff time in intake processing and by reducing the burden on MSFW families. As long as there are separate programs, individual program intake and questioning cannot be completely supplanted, as the Indiana and Iowa experience indicates. But economies could be achieved. Reducing the disparities among the definitions might help encourage the development of consolidated outreach forms, even if separate programs retained some differences in definitions and eligibility criteria that are truly justified by the nature of the particular program.

Moreover, many officials and service providers that we interviewed pointed to another important reason for a uniform definition -- at least a core definition for certain purposes. They reported considerable frustration at being unable to provide legislators or others with an agreed count of migrant farm workers, nor even of the wider category of seasonal farm workers (a problem we explore in detail in Chapter 5). An agreed census or estimation mechanism should help them argue for their budgets, and help them to document trends in the condition of their clients. It would also help to identify the real needs of workers and their families, as well as the appropriate regional or local distribution of funds meant for MSFW services -- a targeting function that is not well served at present.

Finally, we believe that experience with a uniform core definition, even if it were used initially only for population counts or for part of the outreach process, would have beneficial long-term effects. Over time, it may ease the concerns that have in the past sparked resistance to proposed changes in the definitions governing particular programs, and may facilitate incremental progress toward more uniform eligibility criteria.

NOTES

[1]Interviews with Nancy Quynn, Peninsula Legal Services, Eastern Shore of Va. (July 31, 1991); Kevin Boyd, Telamon Corporation, Richmond, Va. (July 31, 1991).

[2]National Migrant Resource Program, Inc., Integration and Coordination of Migrant Health Centers, at III-25 (report submitted to HHS, Feb. 28, 1992).

[3]The information in this section is drawn from an interview with Lynn Clothier, Indiana Health Centers, Inc., Indianapolis (Aug. 16, 1991), and from descriptive literature of the Consolidated Outreach Project.

[4]Interview with Terry Meek, Executive Director, Proteus Employment Opportunities, Inc. (February 28, 1992).

[5]Id. Proteus also does some outreach to farmworkers in connection with the Wagner-Peyser Act, 29 U.S.C. §§ 49-49k (1988).

[6]NMRP, supra note 2, at III-21 - III-36. The study also contains illuminating case studies of eight other programs and offers useful conclusions and recommendations on coordination and integration of services to MSFWs.

[7]Id. at III-24.

[8]Id. at III-25.

[9]A survey by the National Association of State Directors of Migrant Education (NASDME) noted this problem. One respondent commented: "The others look at coordination as 'How much money do you have for us?' or else there is apprehension: 'Do you want *our* money?'" Testimony of Beth Arnow for NASDME, before the National Comm'n on Migrant Education, April 29, 1991, at 2-3.

[10]Much of this problem in Indiana appears to have resulted from a misunderstanding. The consolidated outreach staff saw its task primarily in terms of developing full information on the MSFW population in the state, not as encouraging individuals to enroll in particular programs. This approach served most MSFW programs reasonably well, for they could generally rely on other incentives (like an obvious need for day care or medical treatment) to bring about actual use of their services.

In contrast, JTPA is more dependent on proactive recruiting to persuade individuals to join programs that take them out of the workforce for training. The former JTPA grantee in Indiana, a state agency, waited for the Consolidated Outreach Project to fill this recruitment need, rather than supplementing the more limited consolidated efforts with its own recruitment staff. As a result, the number of enrolled JTPA 402 participants remained quite low, and that JTPA grantee eventually lost the grant. The new JTPA grantee in

Indiana, a private nonprofit agency, has declined to take part in the consolidated outreach program.

[11]A recent poll of MHS grantees, for example, found strong support for the notion that the different definitions impede interagency coordination and for the proposal that a single definition of migrant should be adopted and used by all agencies. Testimony of Frank Fuentes, Chief, Migrant Programs Branch, Administration of Children and Families, Dep't of Health and Human Services, before the National Comm'n on Migrant Education, April 29, 1991, at 15.

[12]A somewhat different version of cross-eligiblity was enacted by the 1992 amendments to the HEP/CAMP program. Applicants may now qualify under the traditional definition, with its 24-month look-back criteria, *or* by meeting the standards of the Migrant Education program or the JTPA 402 program. 20 U.S.C.A. § 1070d-2 (West Supp. 1993). See Chap. 2, n. 58.

[13]One ME state director commented: "All of the migrant-specific programs can coordinate their hearts out, but there still are not enough resources in sight to serve this population unless there is improvement in access to the mainstream programs." Letter from Ronnie E. Glover to Jeffrey Lubbers, ACUS (Feb. 7, 1992), at 5.

[14]One ME state director concluded: "Within the existing statutory/regulatory framework, the degree of effective coordination possible seems to be limited only by the initiative, energy and good will of the service providers at the state and local levels." Arnow testimony, supra note 9, at 2. See also testimony of Lynda Diane Mull, Association of Farmworker Opportunity Programs, before the National Comm'n on Migrant Education, April 29, 1991, at 7 ("personality conflicts can sometimes be translated into policies that discourage productive coordination.").

4

National Coordination

Existing National Coordination Efforts

Several bodies have provided a measure of coordination among MSFW service programs at the national level. They have evolved over time, and they are sometimes known by different names. This section describes the three principal efforts of recent years.

The Interagency Committee on Migrants

Since approximately 1985, the major migrant service programs have cooperated under the framework of an umbrella committee to provide better information-sharing and coordination at the headquarters level.[1] Currently known as the Interagency Committee on Migrants, it meets quarterly, usually in Washington, D.C. The various federal agencies involved rotate responsibility for hosting and chairing the meeting and setting the agenda. The group publishes a directory setting forth the names, addresses, and phone numbers of specific individuals in the various departments who are involved with the committee, and the list is updated roughly every six months. The directory is useful not only for notifying interested parties of meetings, but also to help contact precisely the right office or official between meetings if an interagency issue arises. (We reprint the latest such directory in Appendix B.)

The Committee includes among its number not only representatives from the "Big Four" service programs, but also a large number of people from other offices, both within the same Departments as those four programs (Labor, HHS, Education) and from elsewhere (Agriculture, Justice, Environmental Protection Agency). Many of the offices represented are not migrant-specific or even farm worker-specific, though their mandates cover farm workers, at least in part. Examples are the Occupational Safety and Health Administration (Labor), the Food Stamp Program (Agriculture), or HHS's Office of Civil Rights. The June 1991 directory contains 68 names from these six departments or agencies, plus a half dozen "other governmental" names.[2] The level of participation in the quarterly meetings varies considerably among the offices mentioned in the directory, and some do not attend regularly.

In addition, nongovernmental organizations that are closely involved in migrant issues are also included in the directory, and are invited to attend the Committee meetings. There they are allowed to take part, although usually at the end of the presentations or discussions by the federal governmental members. These organizations include advocacy and watchdog organizations like the Migrant Legal Action Program or the Farmworker Justice Fund; umbrella organizations for the private nonprofits that are the usual grantees for the major service programs, such as the Association of Farmworker Opportunity Programs (AFOP, which represents JTPA 402 grantees), or the National Association of Community Health Centers (which represents Migrant Health clinics); and a few major nonprofits which are themselves grantees, such as the East Coast Migrant Head Start Program. The nonprofits have been pressing for a larger role and recently proposed to take responsibility for hosting and chairing a meeting, but this initiative has not yet been accepted.

The meeting agenda may be built around a specific issue of interest to several organizations, or it may center on a presentation by an invited guest, for example, someone who recently completed research bearing on migrant farm workers. Usually it also includes some time for updating reports on recent initiatives of the major agencies that attend. Those who participate agree that the major function served by the Committee is information-sharing; it is not a policy-making body. For this reason, several people interviewed expressed frustration or impatience with it, and they noted that agencies often tend to send rather low-level personnel who cannot commit the agency but serve instead to report back to policy-level officials. Moreover, even if the lead officials of the migrant programs attend, they are sometimes hampered in assuring real policy changes for coordination purposes. As one of these officials pointed out during an interview, he is six layers below the Cabinet Secretary. Even if he and a counterpart at another agency agree that some mutual change would be beneficial, both will probably have to arrange for their two secretaries to reach agreement before action can be taken. Despite these problems, most who take part agree that on balance the Committee is useful, even if it has limited promise to generate significant program changes in the interests of coordination.[3]

The Farmworker Interagency Coordinating Council

Frustration at the slow pace or modest ambitions of the Interagency Committee helped spur the creation of a second coordination forum beginning in mid-1990. Led by the efforts of John Florez, then Deputy Assistant Secretary in the Employment and Training Administration at the Department of Labor, the key agencies participated in an ad hoc group that came to be known as the Farmworker Interagency Coordinating Council. This was intended to be a meeting of policy-makers, at the Deputy Assistant Secretary level or higher, to focus on particular issues that require policy resolution, and specifically to look for new and creative ways, transcending agency parochialism, to serve migrant families comprehensively.[4]

Members of the Council, as of April 1991, were Labor Department offices with migrant responsibilities, Migrant Head Start, Migrant Education, Migrant Health, and the Department of Agriculture.[5] Mr. Florez left the Department of Labor in 1991, before the Council had time to demonstrate any significant fruits of its labors. It is not clear what will become of this body, although it is apparently inactive at the present time.

The Migrant Inter-Association Coordinating Committee and Coalition

The grantee service providers in the various MSFW service programs gained some acquaintance with each other's perspectives through their participation in the Interagency Committee. This sharing led to a proposal in 1989 that the fall quarterly meeting of AFOP, the principal organization for JTPA 402 grantees, be held at the same time and place as that of the Migrant Education program. Because this gathering was regarded as a success, the participants decided to repeat and expand the process. Planning began then for a comprehensive National Joint Conference on Migrant and Seasonal Farmworkers, to include not only JTPA and Migrant Education, but also Migrant Health and Migrant Head Start. The conference, which served as the annual meeting for all four umbrella organizations, took place in April 1991 in Buffalo.[6]

At Buffalo, some 120 panels and programs offered forums for activists, officials, and participants to discuss specific questions of mutual interest, and to learn of successful local programs elsewhere whose ideas they might want to borrow, including ideas about interagency coordination. And of course the gathering provided for an abundance of informal contacts outside the scheduled meetings. The organizing committee also wanted to award special recognition to a body or bodies that had been especially successful in promoting interagency coordination. It proceeded by agreeing on criteria in advance and then receiving and considering nominees from around the country. The award was a highlight of the proceedings. Many persons interviewed felt strongly that the joint conference and associated activities contributed importantly to interagency understanding, and they have high hopes that concrete improvements in coordination at the state and local level will flow, incrementally, from the contacts made and ideas shared at these meetings.

Pleased with the results of the conference, the participants decided to plan for another joint meeting in 1993. The body charged with responsibility to organize that gathering was also asked to consider other initiatives that might be undertaken by the grantee community acting together. At follow-up meetings in Washington in May and in Denver in October 1991, they initiated planning for the exact structure and organization of this new "Inter-Association Coalition," and discussed other specific tasks for the organization. These may include planning smaller scale workshops for state-level personnel, improving the use of existing publications, selecting current legislative issues on which mutual strategies might be adopted, and working to include other associations in the coalition.[7] In the meantime, the planning body has taken the name of Migrant Inter-Association Coordinating Committee, consisting at present of a total of 11 persons representing the four main migrant assistance programs (ME, MH, MHS, and JTPA 402).[8]

Other National Coordinating Initiatives

In addition to these efforts, the central offices of the various programs have entered into cooperative ad hoc arrangements over the years, sometimes enshrined in formal Memoranda of Understanding or similar documents. These tend to focus on specific areas where there is a clear and recognized benefit from close cooperation. For example, Head Start programs have always considered health screening and treatment an important component of their local services, and Migrant Head Start has logically looked to Migrant Health clinics for assistance in fulfilling this mandate, in those parts of the country where both programs are active. In 1984 Migrant Head Start entered into a three-year interagency agreement with Migrant Health, meant to coordinate policies at the national level and to foster working relationships and joint planning among MHS and MH grantees at the local level. Although the agreement has officially lapsed, patterns of cooperation it fostered have continued, and there is some interest in renewing the formal agreement.[9]

Similarly, JTPA 402 programs can provide services for older teenagers who have difficulty in a formal school setting. This fact sets the stage for cooperation with Migrant Education programs (which tend to focus their services in the school systems). Since early 1990, the Office of Migrant Education and its service providers have been meeting with Labor's Division of Seasonal Farm workers and its service providers to find ways to take advantage of these potential commonalties of interest. The resulting "Coordination Workgroup," which has often included participation at high levels from both departments, has developed the framework for a cooperative agreement between the two programs. The agreement would incorporate clearer policy directives to grantees, placing a high priority on local coordination.[10]

Nevertheless, many obstacles stand in the way of wider use of such agreements. First, they cannot override statutory or regulatory requirements of the specific programs, and these technical objections have sometimes delayed conclusion or implementation of agreements for lengthy periods of time. They have also sometimes led to time-consuming semantic disputes over the exact wording of the agreements. Sometimes effective implementation is also hampered because different levels of government or nongovernmental players are the grantees. For example, ME operates through state-level grantees, whereas the other programs tend to focus on local agencies or organizations. Many times it is not clear to participants that their agencies will gain enough from a formalized relationship to make it worth the trouble of negotiating such an arrangement.

Evaluation of National-Level Coordination

Although they usually express the view that coordination is improving, officials and service providers frequently voice dissatisfaction with the current arrangements for national-level coordination. To assess the adequacy of current bodies or mechanisms fully, however, requires clarity about the objectives of coordination at that level. What follows is our effort to distill the principal objectives that are implicit in the evaluations we have heard, or that seem appropriate to add to the list.

Information-Sharing

The most basic starting point for coordination is sharing of information, so that participants in one program have a better idea of the operations and statutory framework of the other programs, as well as the services they provide and the legislative or programmatic issues they are now facing. This was apparently a central interest of the congressional committee that first proposed chartering the National Commission on Migrant Education in 1987. In explaining the tasks of the Commission, the committee suggested that it explore a "National Center for Migrant Affairs to help coordinate and disseminate information pertinent to migrants."[11]

This objective is not controversial, and it is the one that existing arrangements principally serve. Although the federal government now lacks the single central depository for migrant-related studies and information that was apparently contemplated in the proposal for a National Center (to be discussed below), numerous depositories with narrower focus exist, and in any event the interagency and inter-association mechanisms mentioned above accomplish a considerable amount of information sharing. That the function might be done more systematically is certainly possible, but any efforts toward that end should build on an understanding of the other objectives of coordination.

Geographic Targeting of Resources

MSFW-specific service programs have insufficient resources to serve the entire target population. Thus programs should be located where the highest concentrations of migrant and seasonal farm workers can be found. Each individual program has substantial internal incentives to follow this common-sense dictate, and in general, existing programs have conformed to this requirement as they first let contracts or awarded grants and then expanded. The problem is that farm work patterns change, sometimes quite rapidly. Labor-intensive crops in one area may be replaced by others that can be harvested mechanically, or, alternatively, a large agricultural company may start up a major new labor-intensive operation employing thousands of migrants in an area where such workers were previously unknown. The concern that services keep pace with changes in agriculture also figured in Congress's decision to establish the National Commission,[12] and was frequently voiced by officials and service providers in our interviews.

The present service infrastructure is not well equipped to adjust to these changes. Officials in every program interviewed noted their program's deficiencies in this regard.[13]   Although central grant administrators may cut an established grantee's funding at the next renewal if the grant proposal shows reduced "productivity" or otherwise discloses a decreasing population of eligible recipients, they are not well positioned to spot wholly new areas of migrant activity. Particularly because their budgets (with the exception, recently, of Migrant Head Start) have been relatively level for some time, the agencies have had little capacity to entertain applications for new centers or clinics in previously unserved areas.

Targeting could of course be improved without interagency coordination, but tracking geographic shifts in farm worker activity on a joint basis and responding together presents many advantages. Improvements in quarterly or annual farm worker census figures, as discussed in Chapter 5, would facilitate timely program adjustments of the kind envisioned here, and would also make it easier for a central administrator to feel more confident about cutting or eliminating programs in certain areas where productivity has declined (if it can be shown that this is related to a long-term reduction in farm worker population). Detailed data of this kind, with frequent updates, are expensive to gather. It makes sense to pool resources to provide for the most effective single counting process possible.

Even without such improved data, however, the present coordination entities serve the objective of geographic targeting only marginally and incidentally. It is possible that their meetings provide occasions for, say, Migrant Health to learn of a substantial migrant population in a new area, because they hear of substantial new Migrant Education activity there. But the information-sharing provided by current bodies relates primarily to program in existing locations; they incorporate no systematic effort to use the meetings as a basis for these sorts of geographic adjustments. Many service providers and officials interviewed expressed a wish for more systematic information about shifting agricultural labor patterns, so that they could adjust program accordingly.

Minimizing Program Overlap

There is a clear potential for overlap or duplication among the programs, particularly with regard to repetitious intake processing and outreach. As discussed above, this problem can be minimized greatly through local cooperation, but many localities, for a variety of reasons (inattention, personality conflicts, inertia), have simply been unable to work out the necessary arrangements.

The potential for overlap has also expanded in recent years, as statutes or regulations have authorized service providers to reach wider populations -- for example, by expanding the respective age ranges of the programs. Migrant Education now has authority to count and to serve children and youth from age 3 up, and the Department of Labor has been urged to allow its § 402 grantees to engage in "employability enhancement" services that reach children below its traditional limit of age 14. Coordination would seem to be useful to resist the internal pressures for such expansion, or at least to review more systematically, before any such change becomes a *fait accompli*, whether the program objectives that drive such an amendment could actually be served more effectively through program adjustments in another agency that already reaches migrant children in the affected age bracket.

Such changes in program scope are not normally presented to the interagency forums beforehand[14] (although the changes may be the subject of information-sharing after the fact), in part because those forums have no policy-making authority on such matters. Such advance checking would be desirable, but participants often worry that it would only trigger negative reactions based on "turfism." Nevertheless, coordination would be better served by some such review. Indeed, interagency bodies ideally would not only review proposed expansions of a particular program's authority, but also should look systematically at existing overlaps and think creatively about ways to serve the target population more efficiently. Such scrutiny need not always mean selecting one agency over the other as the exclusive provider -- competition or complementarity may be worthwhile in some circumstances, depending on the task and the geographic area[15] -- but any such overlap should be chosen as the result of careful consideration of a wider range of issues, rather than left to proliferate as a result of a dynamic that seems internal to each program, without close attention to effects elsewhere.

Ideally, government coordination mechanisms should also have the capacity to ask larger questions. Of the approximately $600 million that now goes into migrant service programs, over 50 percent goes to Migrant Education. Is this a sensible way to allocate limited resources? What are the relative priorities of the various services in maximizing the welfare of the target population? It may well be that ME deserves exactly this sort of priority, but under the present arrangements these comparative questions are never expressly asked and answered. There is no forum that effectively looks at such budgetary priorities. We were surprised to learn in the course of our interviewing that even the Office of Management and Budget (OMB) is not organized so as to ask questions of this kind. OMB scrutiny of migrant programs is divided departmentally, with different staff specialists overseeing the respective

MSFW service programs. No single officer in OMB (or elsewhere in the executive office) takes a look at the whole MSFW service program landscape, so as to watch for opportunities to make better use of overall resources in meeting the comprehensive needs of farm worker families.[16]

Many service providers and officials objected to our laying too much stress on the objective of reducing or eliminating overlaps or otherwise pruning or reorganizing programs in the service of a supposed efficiency. Although they could not quarrel with these objectives in the abstract, they worried greatly about what they would mean in practice. They fear that prominent discussion of overlaps in authority will be used as a pretext for serious budget cutbacks in one agency, without any guarantee that the funds will go to another agency to assist those people who lose the service from the first. "Outsiders," in other words, might be too ready to seize on such information to undercut programs they have never supported, and "wasteful duplication" is an easy rallying cry that often simply masks misunderstanding of the real tasks involved.[17]

Others we interviewed pointed out vigorously that the overlaps themselves are far more theoretical than real. All MSFW service agencies serve only a fraction of the eligible population. Overlap in authority simply means that some of those left out of one program for lack of resources have a chance for similar services through another. Migrants who arrive in an area after a MHS center has already reached its enrollment limit, for example, may be able to secure some help towards providing child care through ME resources or through the support services component of JTPA. Overlap, these persons suggest, will not be a real problem until all agencies are funded at a level that permits services for virtually all of their target populations. In any event, they argue, a degree of overlap can actually be of benefit, for it allows experimentation with different approaches, rather than a stifling uniformity.[18]

These reservations deserve serious attention. Overlap in authority does not necessarily mean actual duplication or wasteful spending, particularly at the present level of funding. Moreover, glib talk about efficiency sometimes does mask efforts to gut a program. Nevertheless, we believe that these concerns should be heard more as cautionary notes, to be met by fuller airing of both the pros and cons of particular proposals to reduce overlap or reconsider funding priorities. They do not overcome the desirability of more serious and comprehensive attention to issues of efficiency and potential duplication of effort, particularly with regard to the two issues identified above -- the burden of duplicative outreach or intake procedures, and the dynamic of incremental expansion in individual service jurisdictions. At present, decisions about funding and siting of programs, or other expansions of authority, are made separately by the different federal agencies, in processes that are principally responsive to the grantee constituencies of that program. It would be far better to have some forum for examining the overall service package in a given area, in a way that takes full account of the ultimate objectives for service to the MSFW population -- without ignoring, of course, the risk of pretextual cuts that so worries the agencies.

Viewed in this light, current coordination entities are not well-designed to serve these efficiency objectives. They function as a product of comity among agencies or organizations, and none are mandated by statute, regulation, or executive order. Comity could be significantly threatened by proposals, for example, to transfer nearly all outreach staff to a single program so as to consolidate intake processing. It is similarly endangered if one agency begins asking persistent questions about another agency's program expansions, and even more so if the first aggressively suggests shrinkage in the scope of the second agency's mandate on the ground that the first can serve a certain population or meet a particular need more effectively. The present interagency bodies, dependent as they are on continuing goodwill of the participating agencies and lacking any legal requirement that such agencies continue their participation, are unlikely ever to provide a good forum for asking these kinds of tough questions.

Comprehensive Services for Migrants

Most of the agencies that serve MSFWs, and particularly those with an educational focus, recognize the need for comprehensive services if their own objectives are to be fully realized. Only if children are well-nourished and healthy, for example, can they take maximum advantage of their schooling. Workers cannot be steady participants in a job training program if their child-care arrangements are unreliable. Education on good hygiene may be highly useful in preventing future illnesses or injuries and thus minimizing the need for treatment in a migrant clinic. Precisely because of this recognition, most of the agencies have authority to spend some of their funds on ancillary or supportive services. Sometimes these authorities are flexible enough to allow the agencies to fill gaps in the overall program landscape, by providing needed services that are not the specific target of any of the service programs. (An example is transportation services, which can be provided as non-training related support services by JTPA grantees,[19] or, in some circumstances, by ME or MHS.) According to their statutory or regulatory requirements, most are to use their funds for these purposes only after it is determined that no outside agencies can provide the service. This is a difficult mandate to realize in practice, however, even though grantees are required to discuss in their grant applications the general steps they are taking toward these ends.

In addition to this concern for filling gaps in service provision, interest has been renewed in transcending traditional agency boundaries in order to consider and address the needs of migrant families as a whole, perhaps through a case management approach that would give the family a single point of contact within the service provider bureaucracy in any local area.[20] This impulse played an important role in inspiring the establishment in 1990 of the Interagency Coordinating Council.[21] That Council did not remain active long enough to know how effective it might be toward that end. But even if revived in the same form, it is likely to run into many of the same problems discussed in the previous section on program overlap. Lacking ultimate decision making authority, or even a foundation in statute or executive order that mandates continuing participation by affected agencies, the Council too is dependent on persuasion and goodwill to have its suggestions implemented, and indeed to continue functioning at all. The agency-focused, task-specific outlook of the participants thus imposes a barrier to implementation of any agency-transcending ideas that body might generate. It does not have independent authority to initiate even limited pilot projects meant to demonstrate the possibilities for comprehensive approaches.

Toward a Better Coordinating Entity

Harold Seidman has described the quest for coordination as the "twentieth century equivalent of the medieval search for the philosopher's stone. . . . If only we can find the right formula for coordination, we can reconcile the irreconcilable, harmonize competing and wholly divergent interests, overcome irrationalities in our government structures and make hard policy choices to which no one will dissent."[22] Coordination is always attractive in the abstract, yet often painful and difficult in the concrete. The relatively widespread support for coordination among migrant service agencies suggests an equally rosy view of what coordination will accomplish. This is coupled with a tendency to downplay the painful adjustments, including some loss of control, occasional ceding of program responsibilities, or transfer of funding, that complete coordination is likely to entail for at least some of the individual agencies involved. This view may explain why much of the concrete discussion to date focuses on relatively painless issues like better information-sharing or highly technical matters like definitional discontinuities. Definitional differences become a form of excuse, affording an explanation for the failure to take the considerable time required or to make the hard decisions that may be necessary to adjust program so as to achieve real efficiencies or to better serve the overall needs of migrant families.

Objectives

The foregoing discussion suggests that a future interagency coordination entity needs above all to be able to ask creative, persistent, and tough questions about the allocation of responsibilities and the ways in which ultimate service objectives (which transcend agency boundaries) are or are not being served. This need not mean, necessarily, that major changes in program operations are in the offing. It does mean that more careful scrutiny, from a perspective not tied solely to one agency and its constituencies, would be applied -- initially to examine proposed changes or expansions in mandate, and eventually to review existing programs. At least temporary disruptions in agency comity must be possible without terminating the coordination endeavor, although effective coordination over the long run will of course require skill and tact to move beyond such challenging periods with a minimum of lingering bad feeling.

Second, the coordinating entity needs the capacity to engage the attention of the appropriate policy-making officials in the affected agencies, sometimes up to the level of the Cabinet Secretary.[23] If matters cannot be resolved through such channels, the coordinating entity needs the capacity to assure ultimate interagency resolution, if necessary (when the issues are of a scope to warrant this) by means of Presidential choice among competing options.

Third, the entity should also pay attention to ongoing information sharing, and should initiate improvements where possible at reasonable cost. This objective, however, is closer to adequate realization under the present system than are the previous aims.

Fourth, the coordinating body should assume a function related to the preceding: it should oversee a process leading to the development of better statistical systems, as discussed in Chapter 5, so as to provide agencies with agreed and reliable information on MSFW populations, and especially on changes in farm work and in farm worker population patterns.

Fifth, a federal coordinating entity would be the logical focal point for efforts to promote coordination at the state and local levels. It could be the central decision maker in choosing award recipients or providing other recognition for successful local coordination efforts. It should also work to devise other incentives to promote these ends.

Sixth, the coordinating entity should also take the lead in examining possibilities for harmonizing definitions, if only for purposes of establishing a "core definition" to be used in census counts. Over time, it may also find ways to harmonize program definitions or other eligibility qualifications.

Alternative Models to Promote Coordination

In our search for improvements in coordination among MSFW programs, we have not found the philosopher's stone. No one proposal or set of proposals for coordination clearly recommends itself as superior to all others, particularly since a wide spectrum of variations and permutations can be imagined. We have tried to avoid overdoing such detail; instead we cluster the ideas around four possible models. The first is the most modest, addressing only the improvement of access to information, which is not a priority need at present. It could be implemented in conjunction with any of the other three, more comprehensive, options.

A National Center for Migrant Affairs. There have been recurring proposals for creating a central clearinghouse for information on migrant programs, as a way of improving the information-sharing function that is necessary to better coordination. As noted above, this idea received concrete expression in the legislation that created the National Commission on Migrant Education. The statute requires the Commission to consider whether there is "a need to establish a National Center for Migrant Affairs and what are the options for funding such a center."[24] The legislative history describes the purpose of such a proposed center as "to help coordinate and disseminate information pertinent to migrants," making specific reference to a consultant's study that urged consideration of a central repository for "products" generated by the coordination grants of the Migrant Education Program (now known as § 1203 grants).[25]

We encountered no substantial objections to the proposal for a National Center. To the extent that it is simply an information clearinghouse, it steps on no one's programmatic toes, at least in the absence of specific plans for funding it. But we found little enthusiasm for the idea either. The specific concerns of the consultant's study referred to in the House report, which focuses on Migrant Education "products," have been met by the more complete development of three Program Coordination Centers (PCCs), one for each of the western, central, and eastern migrant streams, under the umbrella of the Migrant Education Program. As discussed in Chapter 2, the PCCs, in addition to other functions, serve as repositories for the products of previous § 1203 grants, which can be drawn upon by any interested user.[26]

Many other existing resource centers also can be consulted by those who seek further specific "products" associated with the education of migrant children or the other migrant services. For example, the Department of Education maintains an elaborate and technologically advanced system of resource centers as part of its ERIC system (Educational Resources Information Centers). The staff of the 16 subject-specific ERIC clearinghouses, operated under contract with the Department, review the documents and journals they receive, abstract and index those that are relevant to their center's subject matter, and respond to inquiries from teachers, parents, students, and researchers. ERIC produces monthly hard-copy indexes and quarterly CD-ROM directories, and all the indexed documents can be consulted in microfiche form at any of some 900 depositories throughout the United States and in numerous foreign countries.[27] One of the 16 centers, known as CRESS (Clearinghouse on Rural Education and Small Schools), includes migrants within its coverage, but documents on migrants give rise to only a small fraction of its activities.[28]

Other migrant programs have their own associated resource centers or clearinghouses. For example, Migrant Health funds the National Migrant Resource Program in Austin, Texas, which provides services similar to ERIC's for migrant health issues. Migrant Health clinics make use of its database and library, as well as the other specific products (such as medical protocols) it generates. Migrant Head Start agencies and staff similarly draw upon the services of the Migrant Head Start Resource Center in Tysons Corner, Virginia.

With all these repositories in existence, it is hard to develop a persuasive case for adding still another. The problem is not the lack of centers capable of disseminating available information on migrant service programs.[29] The problem is more the disconnection between the existing repositories and the field-level service providers. Although the latter might well benefit from learning of relevant studies or accounts of strategies devised elsewhere to overcome problems similar to ones they are now facing, few local service providers have the time to engage in this kind of research and reading. Further consolidation of migrant-related information into a single National Center is unlikely to ease this situation. If wider or more effective use of such "products" is deemed a priority, any extra funding might well be better used instead to beef up the staffs at the local level, in the hope that some of the extra staff time might be devoted to drawing upon such accumulated learning.[30]

We recommend that any effort to create a National Center for Migrant Affairs not start from scratch. It should instead build on the existing foundation of current documentation and research centers.

Most promising (and least expensive) would be some form of loose linkage among existing clearinghouses under a National Center's umbrella, without greatly disturbing their present operation. The National Center would then focus its efforts on publicizing the resources available, easing the use of such resources by local service providers, and perhaps providing a central telephone exchange which would refer inquirers to the appropriate clearinghouse. It should be a modest undertaking if undertaken at all. Care should be taken to avoid duplicating existing resources or draining funds from other direct services.

A Separate Department or Agency. One can envision the creation of a single Migrant Affairs Department that would unify all migrant service functions. A model for such unification might be found in the creation of the Department of Energy in 1977. That reorganization was advocated, in significant part, on the ground that it would provide for better coordination among programs that had previously resided in separate agencies.[31] The new department combined functions that had previously been performed by five separate departments and four independent agencies.[32]

A new Department of Migrant Affairs could incorporate the Big Four programs along with others such as migrant housing from Agriculture, and possibly even the authority to serve MSFWs now carried by general programs like Food Stamps or WIC. (It is possible also, in the interest of true comprehensiveness, to envision an agency that would also eventually incorporate enforcement functions like those established by the Migrant and Seasonal Agricultural Workers Protection Act.[33]) Coordination among programs would still be required, but it would become an intra-agency task, and could more readily command attention at the policy-making level. The process of preparing the agency's annual budget, plus defending it as a whole before OMB and the Congress, would provide built-in occasions for asking many of the comparative questions mentioned above. Beyond the discipline necessarily imposed by the budget process within a single department, a departmental policy planning bureau might assume ongoing responsibility for asking tough questions about overlaps and efficiency, with authority to report its suggestions directly to the Secretary, along with comments by the affected programs.

Governors and local officials would know where to concentrate their efforts if they want to lobby to reduce certain duplicative procedures. In addition, a unified department of this type could ultimately develop a unified grant process, inviting combined proposals by a single service provider for each local area or region that would incorporate all the components now handled separately -- e.g., education, job training, nutrition, child care and child development, health services, housing, etc.[34] Applications could be judged as to how well they mix and match the various components so as best to meet the comprehensive needs of migrant families in the local area, in light of the local or state resources, public and private, already available.

Obvious obstacles stand in the way of realizing this proposal in practice, and not even those who have been most supportive of ambitious coordination efforts or of consistent Cabinet-level attention to migrant issues have seriously advanced the idea of a new Department of this kind. One initial objection derives from size and relative political priority. Migrant service programs now claim a combined budget of something over one-half billion dollars, whereas the Department of Energy receives $ 14 billion (and also touches on issues that doubtless have wider political ramifications).[35]

Unification of all MSFW service programs could still be accomplished in other ways, even if a new Department is deemed inappropriate. The Migrant Affairs body could instead be set up as a non-Cabinet agency, something like the Small Business Administration.[36] Alternatively, the unified body could be designated a bureau, headed by an Assistant Secretary, and located within one of the existing departments, although it is by no means clear which department should be awarded such functions.[37] These modifications would create a structure more proportionate to the scope of MSFW service programs, but either one would generate considerable political controversy over the exact institutional arrangements or the precise forms of accountability to President and Congress.[38]

A more fundamental objection to this sort of unification is the following: what might be gained in coordination among migrant-specific programs would be outweighed by the losses to effective coordination with other related governmental functions.[39] Migrant Health, for example, would have to find new ways to assure interagency coordination with other public health programs, including Community Health Centers, and Migrant Head Start would probably need to coordinate, now across agency boundaries, with the basic Head Start program. These needs could require massive readjustments. Furthermore, some people we interviewed saw the current departmental locations of these programs as a form of mainstreaming, helping to minimize the isolation or stereotyping that sometimes befalls the programs' clientele. Finally, consolidation into one agency might make all the programs more vulnerable at budget time.

The point need not be labored. Gains in efficiency and coordination under this model would probably be overshadowed by these other disadvantages. Complete unification is not a practical option. Instead it might be viewed largely as an ideal type, illustrating the maximum effort that might be made if efficiency and coordination (or more accurately, integration) assumed the highest priority. It provides a kind of lodestar that might illuminate other options that are more realistic in the medium term.

Improved Interagency Council. Interagency committees and councils are often used for coordination, but they can hardly claim to have gained popularity or even respect. Seidman summed up common attitudes in referring to them as "the crabgrass in the garden of government.... Nobody wants them, but everyone has them. Committees seem to thrive on scorn and ridicule."[40] President Carter particularly targeted such committees as part of his effort to streamline and rationalize the organization of the federal government. But he too found that he could not live without them, creating at least seven new councils in one twelve-month period, including an Interagency Coordinating Council to deal with urban and regional policy, an Energy Coordinating Committee, a Management Improvement Council, and a Consumer Affairs Council.[41] Despite their nominal unpopularity, interagency councils obviously meet real needs, particularly when other considerations preclude full integration of the programs involved.[42]

Coordination among the MSFW programs could therefore be improved by strengthening or redesigning the current committees or councils, leaving the basic responsibilities for program in the various departments, under the ultimate responsibility of the respective Cabinet secretaries. To make this approach successful, careful changes in the coordinating bodies would be needed, however, to overcome the deficiencies noted earlier. Changes should address the following objectives: to give the body or bodies higher standing, to assure more complete involvement by officials with policy-making authority, to provide the coordinating entities both the capacity and the incentives to look closely at proposed improvements even if the proposals provoke resistance on the part of one or more of the participating agencies, and to equip them with the ability to force decisions by agency or department heads when a proposal has been sufficiently refined and deserves a straightforward decision.

These objectives are rather easy to list. Mechanisms or procedures for accomplishing them are far harder to craft, because in the end their successful achievement may depend much more on political support or substantive priorities than they do on procedural fixes.[43] In fact, the current mechanisms could be made to work for these purposes, without much procedural or structural tinkering, if, say, the Secretary or Deputy Secretary of one of the Departments involved began to take substantial personal interest in migrant programs generally or in some specific coordination initiative (such as consolidated outreach and intake processing). Similar results might be made to flow from the present structure if the President or a key White House staffer placed equivalent priority on the same matters. What is needed, in short, is someone in a position of sufficient prominence who would press the matter, demand high-level attention in counterpart agencies, spend a fair amount of personal time (a precious and scarce commodity) and political capital, and ultimately refuse to let the issue drop until some resolution is reached -- either implementation or agreed abandonment of the initiative.

Such a scenario is unlikely in the near term, in view of the host of other issues, with higher political salience, that compete for attention of those high-level officials who could energize the existing coordination mechanisms. Procedural or structural changes cannot by themselves provide this sort of impetus. The best they can do is improve the odds that a high-level figure might choose a migrant or farm worker issue for priority attention because he or she knows that there exists a workable forum for refining ideas, implementing change, and eventually (if the changes work out as expected) pointing to concrete achievement.

With these cautions in mind, structural suggestions can be offered that might provide some progress, without displacing the ultimate authorities of the department heads. First, a more solid and enduring basis for the interagency mechanism would help give it higher stature. Some officials and service providers offered in their interviews preliminary ideas for grounding the mechanism in statute or Executive Order. Such a chartering instrument would give the entity authority and a mission transcending the temporary acceptance or acquiescence of agency or department heads. Its drafting would also force all participants to give close thought to just what kind of a body might be most useful.

We recommend, if this option be pursued, that the charter provide for a coordinating entity built generally on the model of the Interagency Coordinating Council, rather than the Interagency Committee on Migrants. The former is a smaller body, meant to pull together a few key people with policy authority for actual decisions on changes in operation. Its smaller size should promote focused dialogue on such issues rather than having the function degrade into mere information-sharing. The instrument should specify exactly which officers would be the members of the Council from the various participating agencies. In general, they should hold the rank of Deputy Assistant Secretary or above; perhaps a higher rank would be appropriate. Designation in statute or executive order of the officers who are to be members cannot, of course, guarantee attendance by the principal rather than a delegate. But express designation does provide a fulcrum for pressure by the chair if one of the named officials is too often absent.[44]

It might also be useful for the charter to tap specific officers (by position, not name), from the departments most heavily involved, as chair and deputy chair, in order to assure consistency and continuity. Such a permanent assignment, however, may prove too rigid. It would in any event be highly contentious, perhaps leaving those agencies not chosen suspicious about the new body from the start. If so, a rotating chairmanship may become necessary, but the rotations should be widely spaced, at intervals of no less than two years, to allow for both some sense of "ownership" in initiatives generated during the period and enough continuity to see many of them through to completion during one agency's chairmanship. This arrangement would also help assure that Council initiatives would be adequately staffed. Realistically the chairing agency will have to staff the process any time others cannot be persuaded to take the lead in preparing reports or otherwise supporting an initiative.[45] Finally, a relatively long-term chair may be in a better position to press other agencies at the highest levels, to assure that they take some action on well-formed proposals and not simply let matters drift in the hope that proponents will lose interest.

The chartering instrument should require advance review by the Council of any significant MSFW service program changes (such as amended regulations or substantial alteration in grantee performance standards) well before their adoption. Adoption would still be within the authority of the originating department, but the Council might well offer an interagency perspective that would otherwise be lacking in the internal deliberations. Sometimes, as mentioned above, this perspective could be expected to help counter certain expansionist tendencies internal to the separate programs. The charter should also assign some sort of comprehensive budget review authority to the Council, to provide a forum for considering comparative effectiveness and deciding whether there should perhaps be a different assignment of resources.[46] Obviously all of these review functions will need to be handled in close cooperation with the Office of Management and Budget. In this connection (indeed, under any conceivable coordination option), OMB should also designate a single office or staff specialist to provide comprehensive budget and regulatory review of all MSFW programs.

Coordination of this type is most likely to be successful when focused on specific tasks. The council could come up with its own agenda under very general terms set forth in its charter, of course. But it is probably better if the chartering instrument itself assigned the council not only the general mandate of coordination but also a few such specific assignments, which would then come with the imprimatur and mandate of Congress or the President. Such an assignment would help build momentum, from the earliest days, for an active agenda for the new coordinating entity. Also, if the body has a specific duty to come up with a definite product in the form of new regulations or other program instruments, all affected agencies will have a definite incentive to remain engaged in the process at a fairly high level, if only to protect their own bureaucratic or programmatic interests.

The most promising early tasks we have identified (ones that need not have terribly threatening short-term consequences for any of the participating agencies) are discussed in Chapters 3 and 5: development of a core definition and detailed plans for improved MSFW census information, creation of a consolidated intake form and a streamlined outreach procedure, and possible local pilot programs allowing cross-eligibility or definitional waivers. The chairing agency can be expected to take the lead on many of the council's projects, but on some issues it may make sense for another agency with a greater stake to be designated as lead agency. In any case, the clear assignment to one of the participating agencies of responsibility for a concrete action outcome, to be reached after consultation with the other agencies, appears most promising for sustaining attention and involvement.[47]

The Interagency Committee on Migrants need not be displaced under this model. By and large, it meets a different need: providing a forum for quarterly information sharing and notification, involving a larger number of operational level staff. Similarly the Interassociation Committee could continue to meet the distinct coordination needs of the grantees of the various agencies. When either of these bodies generates concrete proposals for coordination or altered functioning, the proposal could go to the Council for further consideration and for ultimate implementation.

This option, creating a strengthened interagency council as the principal federal coordination entity, holds certain advantages. It would be relatively inexpensive, and it could be established without major disruptions in familiar agency operations. Some initial wrangling could be expected over the exact provisions of any new statute or executive order, but the chartering instrument could probably be drafted (particularly with a rotating chairmanship) so as to avoid deep disaffection in any quarter early in the process.

The disadvantages are straightforward. Despite any structural improvements that might be devised, the council will still be highly dependent on agency goodwill for its effectiveness. Lacking staff, even the initial development of its proposed initiatives would be dependent on the relative priority assigned to coordination by the participating agencies. The council may not prove to have the clout or sustained interest needed to assure continued involvement by policy-level officers or to force migrant coordination issues onto the agendas of the department heads who would retain ultimate authority.

A Migrant Coordinator or Czar. Many of the disadvantages associated with the interagency council model could be remedied by creating a new central office of coordinator, with an identity separate from any of the constituent agencies and with its own staff. Its mission would focus wholly on coordination of programs and more effective use of overall resources devoted to MSFW service programs. The person appointed as coordinator could also function, within the government and in relations with the public, as a spokesperson and farm worker advocate.

There are numerous possible models for such an office. Three are examined here: the Office of National Drug Control Policy, the Office of Science and Technology Policy, and the Coordinator for Refugee Affairs.

In 1988, after earlier bills had been vetoed, Congress finally succeeded in enacting a statute creating the Office of National Drug Control Policy in the Executive Office of the President.[48] The statute was born of congressional frustration over the perceived inability of the dozens of agencies involved in drug control to come together to develop a consolidated strategy to achieve the nation's objectives, and the new structure replaced three earlier drug policy councils or boards.[49] The office is headed by a director, appointed by the President with the advice and consent of the Senate, and three other top officials are likewise made Presidential appointees subject to Senate approval.

The director is charged with developing, modifying, and insuring the implementation of a national drug control strategy.[50] To this end all other agencies involved in drug control must give advance notice to the Office of any proposed changes in policies. The Director then reviews the proposal and certifies whether it is consistent with the national strategy.[51] He may not directly countermand such policies, however, even if he finds them inconsistent with the strategy. To that extent he does not have genuine supervisory authority -- a hard-fought concession won by the other agencies during congressional deliberations on the bill. The Director has a similar role in budget review of each of the agencies' requests, with the responsibility to certify the adequacy of the budget in light of the published national strategy.[52] (The total budget for these purposes in 1992 is $12.7 billion.) In sum, the director does not have direct authority over agency policies or budgets, but can use the certification process and attendant publicity to "shame" an agency into changing a policy or a budget request, or else force the matter onto the President's desk for resolution. This can of course be a considerable power, but it is dependent on the Director's own skills and relationship with the President. The position was relatively powerful during the tenure of its first incumbent, William Bennett, but was later regarded as considerably weaker under Bob Martinez.[53] Moreover, despite its limited statutory authority, the Office grew to be quite large, with a staff of over 140 and an annual budget of $19 million. Many observers questioned whether it effectively accomplished the coordination objective Congress originally had in mind and the Clinton administration has announced that it will radically reduce the office's staff and operations.[54]

The Office of Science and Technology Policy (OSTP) was created by statute in 1976,[55] to provide comprehensive overview of, and coordination among, scientific endeavors throughout the federal government. The office was placed in the White House, as one of the 11 agencies that make up the Executive Office of the President. As of 1988, it had a total full-time staff of 31, including 12 detailed to OSTP from the other agencies of the government most involved in science and technology questions.[56] The director, who is also the President's Science Advisor, chairs the White House Science Council, an advisory body, and also a Federal Coordinating Council on Science, Engineering and Technology. The Coordinating Council's other members are the senior science and engineering executives of each agency with substantial technical involvement. Other interagency committees on specific scientific fields or on specific new initiatives report to this Council.[57] OSTP helps devise strategies for the most effective use of federal scientific resources, such as in supporting new technologies like superconductors. As described by the director, the system collects a wide range of options and a wide range of views on those options, and then, "while forming consensus where that is possible, highlights issues and ... raises them to a higher level for resolution."[58] It also has a defined role in budget guidance and review.[59]

Although the two previous descriptions suggest several functions like those that might profitably be performed by a MSFW Coordinator, each model's relevance may be limited. Science policy has a much wider reach than do migrant programs, touching on many more agencies and overall policy objectives. Drug control is a major priority for the nation, and the billions of dollars involved in the anti-drug effort dwarf the scope of MSFW assistance.

A model on a somewhat more modest scale, which may be more instructive for present purposes, is the office of the U.S. Coordinator for Refugee Affairs. Refugee admission programs involve the efforts of the State Department, the Justice Department's Immigration and Naturalization Service, HHS's Office of Refugee Resettlement, and sometimes other agencies. After resettlement, other departments, such as Education and Labor, frequently play a role in supporting a successful transition to life in a new homeland. When refugee programs expanded greatly in the late 1970s as the exodus from Southeast Asia accelerated, the need for better coordination among the agencies became acute. Accordingly the executive branch created the office of the U.S. Coordinator for Refugee Affairs in 1979. A year later, when the Refugee Act of 1980 was passed, the office was given statutory mandate and a specific list of duties.[60]

Under that statute, the Coordinator is appointed by the President with the Senate's advice and consent. He or she is responsible to the President for development of overall policy on refugee admission and resettlement, and for coordination of programs "in a manner that assures that policy objectives are met in a timely fashion." The Coordinator is also to design an overall budget strategy on these matters, to "provide guidance" to the agencies in preparing their own budget requests and to give OMB an overview of these issues. The statute specifically mentions the Coordinator's role as an advisor to the three department heads most closely involved, and it also instructs him or her to develop effective liaison with state and local governments and nongovernmental organizations involved with refugee resettlement.[61] The Coordinator, whose office is based in the State Department (an arrangement that has been criticized[62]) holds the rank of ambassador-at-large and also assumes some responsibility for international representation and negotiation. The office now has a professional staff of seven who cover both domestic and international issues and maintain liaison with each of the affected agencies. In earlier years the number of staff has sometimes been much higher, including several officers detailed from other agencies.[63]

Many of the functions of the Refugee Coordinator's office listed in the statute are exactly those that a Coordinator for Migrant and Seasonal Farm worker Service Programs should accomplish. If this option is chosen, however, it would probably be advisable not to locate the MSFW Coordinator's office in any of the departments now having migrant service programs. Whatever the merits of placing the Refugee Coordinator in the State Department,[64] it appears especially important for the MSFW Coordinator to develop a perspective that transcends departmental parochialism. That process would be difficult if staff and budget are under the ultimate control of just one of the departments. Equally important, a coordinator must not be *perceived* as simply an advocate for one of the agencies.[65] The logical place for such a transdepartmental office is the White House (as is the case with the Drug Control Office and OSTP.)

This option could be combined with some sort of improved interagency council, as described in the preceding section. But it holds certain advantages over the former. Giving one individual focused responsibility for interagency coordination, rather than assigning it to a collegial body made up of persons with other substantial responsibilities in their own departments, obviously improves the odds that coordination will receive sustained attention. Unlike a temporary chairman drawn from Deputy Assistant Secretary ranks, the Coordinator may develop the stature to keep coordination proposals on the front burner and before the necessary high-level officials in other agencies, rather than having to work a proposal up through multiple departmental layers and then wait while it gets placed onto the crowded agenda of a department head. The new position would provide a visible sign of the priority given to interagency coordination and the development of improved service delivery to the MSFW population. (On the other hand, recent disaffection with the Drug Control Policy office indicates that such outcomes are by no means assured.)

We encountered two types of strong objections when we mentioned early versions of this proposal to create a federal coordinating entity during our interviews. The first was aptly captured by one official's comment: "Oh Lord, not another entity!" In his view, and that of others who share this opinion, adding such an office simply means proliferating committee meetings and paperwork. Drawing on his own recent experience in another agency that had been brought under the umbrella of the White House Drug Control Office in 1988, he suggested that massive studies might be ordered and strategy papers written, but actual program changes in the interest of improved coordination or service delivery would prove no easier to achieve than at present. Another layer of unproductive bureaucracy, necessarily distant from actual service provision and hence unfamiliar with the real challenges as experienced by field-level providers, would spring into being, competing for resources that should instead go directly for services.

The other set of objections focused on placing this office in the White House. Such a step, some argued, would damage the programs by politicizing MSFW service issues. MSFW issues can of course be highly contentious and adversarial, particularly regarding enforcement regimes like pesticide regulations, housing code compliance, or minimum wage obligations. In those settings, the struggle between growers and farm worker advocates sometimes achieves epic proportions. But services for farm workers have become relatively depoliticized. More and more, agricultural interests see these services as part of the benefits available to their workers, and so are less inclined to resist their extension. They may even become advocates of new clinics or centers for their locality. This state of affairs should not be disturbed. According to those who advance this objection, getting the White House directly involved by means of a Coordinator's office would invite dragging service programs into overtly political initiatives aimed at electoral advantage.

Both these objections carry weight and identify real risks. Whether the risks are worth running depends on an assessment of the gains to be expected from this type of coordination regime. Such gain is inevitably hard to gauge in advance. Its realization depends not only on structural design but also on present imponderables, such as the diligence, persistence, and common sense of the people who fill the key positions. In any event, our interviews did generate a few ideas that might help alleviate some of the concerns reflected in these objections. For example, the office itself should remain small, with a lean staffing pattern; extra staffers mean more capacity to demand reports or other paperwork from the agencies, with less regard for the real need for such endeavors. Perhaps a half-dozen professionals should be adequate, and some of these should be persons detailed from the agencies.

Steps were also suggested to minimize the risk of unseemly political misuse of a coordinator's office.[66] First, to minimize the incentive for some such distortion, the statute, order or memorandum establishing the office should rigidly provide that the coordinator would have responsibility for MSFW service programs only, not for enforcement regimes affecting farm workers. As noted above, the latter regimes are more closely associated with adversarial struggles between growers and workers. Any perception of the office as having significant authority over such enforcement would raise the stakes surrounding the choice of the coordinator, thus increasing the risk of political interventions in a manner that might be detrimental to central service program objectives.

Second, if the office were to be created by statute, the statute itself could specify the coordinator's qualifications and the office's functions, and perhaps include other safeguards minimizing the risk of politicization. For example, as one MHS grantee suggested, the statute might require that the Coordinator be a person with a stated minimum of field experience in MSFW service programs.[67] Although newcomers to the area might well serve with distinction (as has sometimes been the case with the Refugee Coordinator position), requiring such field experience would improve the odds that the office is filled by someone sympathetic to the programs' service objectives and more willing to resist partisan pressures from elsewhere in the White House. Such a person might also be more careful to shield service programs from burdensome reporting or paperwork.

Promoting Coordination:
An Interagency Coordinating Council

Although the question is a close one, we favor the interagency coordinating council model. It is true that many service providers we interviewed seemed excited about the idea of a White House coordinator, but that excitement probably had more to do with the symbolism of such an office than with sound judgments about what it could realistically hope to accomplish.[68] Mere location in the White House is not likely to lead to the priority for MSFW service issues that these providers may expect.

Incremental improvement to coordination through a revitalized interagency council appears more promising and more realistic. Though the most recent effort of this type has foundered, we hold out modest hope that the process can be energized by the sort of institutional charter sketched above, provided by Congress or the President, particularly one that mandates action within a stated time on one or two initial coordination tasks. Although a statutory charter would perhaps be more enduring, it is also more difficult to secure than a document issued by the President. We therefore recommend that efforts should focus on the issuance of an Executive Order containing a charter for an interagency council. The modesty of this conclusion may seem anticlimactic, but we believe that this approach can be made to produce, in a relatively short time frame, some real improvements in the services provided to the nation's farm workers.

NOTES

[1]See testimony of Sonia M. Leon Reig, Dep't of Health and Human Services, before the National Comm'n on Migrant Education, April 29, 1991, at 7.

[2]Interagency Committee On Migrants, June 12, 1991 Mailing List.

[3]See, e.g., testimony of Lynda Diane Mull , Association of Farmworker Opportunity Programs, before the National Comm'n on Migrant Education, April 29, 1991, at 8 ("The influence or impact that this effort [the Interagency Committee] has had is very difficult to measure, and therefore, I cannot say that these meetings have stimulated meaningful coordination of services at the state or local level. However, I do believe the real benefit has been the education that has taken place with Federal agency personnel and other meeting participants at the national level."); Reig testimony, supra note 1, at 7 ("The Committee, in my estimation, has had limited success;" she then lists helpful initiatives, mainly in the realm of promoting better understanding and reducing isolation or fragmentation).

[4]See generally testimony of John Florez, Employment and Training Admin., Dep't of Labor, before the National Comm'n on Migrant Education, Buffalo, April 29, 1991, at 9.

[5]Testimony of Frank Fuentes, Chief, Migrant Programs Branch, Administration of Children and Families, Dep't of Health and Human Services, before the National Comm'n on Migrant Education, April 29, 1991, at 15.

[6]Interview with Diane Mull, AFOP (September 17, 1991); interview with Dan Cardenas, NACHC (October 21, 1991). Federal officials were invited to participate, but they were not officially part of the organizing

committee, nor did the federal government directly fund the Buffalo gathering. The grantee organizations deliberately decided to proceed this way, in order to allow a proper level of federal involvement but to leave the decision process to the grantee community.

[7]Migrant Inter-Association Coordinating Committee, Summary of Meeting May 23-24, 1991, Washington, D.C.

[8]Migrant Inter-Association Coordinating Committee, Meeting Documents, Denver, Oct. 3-4, 1991.

[9]Fuentes testimony, supra note 5, at 17. Several other specific examples of interagency coordination involving MH grantees can be found in the case studies reported in the NMRP study, National Migrant Resource Program, Inc., Integration and Coordination of Migrant Health Centers (rept. submitted to HHS, Feb. 28, 1992).

[10]Mull testimony, supra note 3, at 7-9. There are other examples as well, such as a 1989 Memorandum of Understanding between the Department of Education and Agriculture's Food and Nutrition Service, providing for assistance from local ME personnel in encouraging families to obtain and keep with them the Verification of Certification card provided by the WIC program (Supplemental Food Program for Women, Infants and Children). This card helps the bearer qualify more quickly for benefits at a new location. Migrant participation in WIC increased 12 percent from 1989 to 1990, and a WIC official credits much of this improvement to the interagency cooperation. Testimony of Robert Mulvey, Food and Nutrition Service, USDA, before the National Comm'n on Migrant Education, April 29, 1991.

[11]H.R. Rep. No. 95, 100th Cong., 1st Sess. 38-39 (1987). The statute specifically charges the Commission with responsibility to consider such a Center. 20 U.S.C. § 2839(c)(12) (1988).

[12]H.R. Rep. 95, at 38: "Among other things, the Commission will examine the changing demographics of the migrant student population in an effort to assure that the patterns of migrancy are anticipated and the children are served to the best extent possible."

[13]The question arises in somewhat different fashion for Migrant Education, which deals principally with state-level governmental agencies, who are then responsible for using the resources provided to the state in a way that maximizes their effectiveness. The problem of geographic targeting still arises (and the ultimate effectiveness of the programs would still be greatly aided by better systems for reliable MSFW population data), but the responsibility for adjusting program falls mainly on state program officials, not federal officials.

[14]Diane Mull, Executive Director of AFOP, testified: "In some cases, the overlap in program's age ranges is not a complication due to the

different scope of services being offered by each program [but potential] for complications is created when changes occur without consulting other affected service providers. ... [D]iscussions and hearings, like those being held today, should take place before such changes are implemented." Mull testimony, supra note 3, at 2.

[15]See generally Martin Landau, Redundancy, Rationality, and the Problem of Duplication and Overlap, 29 Pub. Admin. Rev. 346 (1969).

[16]This situation differs from, e.g., the review of refugee programs in the days before creation of the office of the U.S. Coordinator for Refugee Affairs. At that time, refugee resettlement programs involved the State Department's Office of Refugee and Migration Affairs, the Justice Department's Immigration and Naturalization Service, and HEW's Office of Refugee Resettlement. OMB officials frequently convened meetings of officers from all three departments to consider relative priorities, to examine ways in which one agency's decisions affected the budget of the others, and occasionally to decide whether a function could be more efficiently handled by another.

[17]See James Q. Wilson, Bureaucracy: What Government Agencies Do and Why They Do It 265 (1989).

[18]See Landau, supra note 15, at 354-56.

[19]20 C.F.R. § 633.304(b)(2), (c)(4) (1992).

[20]This objective has been articulated since at least the days when migrant programs were operated by the Office of Economic Opportunity (See Sar A. Levitan, The Great Society's Poor Law: A New Approach to Poverty 250 (1969) (quoting OEO planning document)), but it has always been imperfectly realized. Moreover, the impulse to provide comprehensive services to families, with a minimum of bureaucratic confusion, is finding expression on many fronts, and Congress has authorized various pilot projects. If these result in promising new initiatives, migrant programs should of course employ the new approaches. See, e.g., 56 Fed. Reg. 29,656 (1991) (HHS request for proposals to establish a National Service Integration Resource Center and announcement of funding for up to six local or regional facilitators for comprehensive integration).

[21]See generally Florez testimony, supra note 4, at 16-17.

[22]Harold Seidman & Robert Gilmour, Politics, Position, and Power 219 (4th ed. 1986).

[23]This conclusion assumes, of course, that MSFW service programs continue to be based in separate departments. We expect this to be the case, but we discuss below one option for total consolidation.

[24]20 U.S.C. § 2839(c)(12) (1988), enacted by Pub. L. No. 100-297, § 1439, 102 Stat. 193 (1988).

[25]H.R. Rep. No. 95, 100th Cong., 1st Sess. 39 (1987). The consultant's study, as described by the House committee report, noted the absence of a "central repository . . . for products associated with the Section 143 grants (coordination of migrant activities) programs." In this respect, the suggestion in the statute, for an overarching center for migrant affairs, goes considerably beyond what the study seemed to have in mind. Section 143 is now referred to as § 1203 (codified at 20 U.S.C. § 2783 (1988)), but the substance is not substantially changed; the provision focuses on interstate and intrastate coordination *within* Migrant Education -- i.e., among ME agencies at the State and local level. Moreover, the full study points out that such products tend to deal less with coordination as such and more with development of curriculum guides, staff training packages, and similar items -- all items fairly specific to ME and probably not of major interest to other MSFW service organizations. N. Adelman & C. Cleland, Descriptive Study of the Migrant Education Section 143 Interstate and Intrastate Coordination Program 41-45 (Policy Studies Associates, Inc., March 1987).

[26]Interview with Saundra Bryant, Office of Migrant Education, Washington, D.C. (October 21, 1991).

[27]Interview with Robert Stonehill and Pat Coulter, ERIC offices, Washington, D.C. (October 22, 1991); A Pocket Guide to ERIC (April 1991); ERIC Annual Report 1991; Directory of ERIC Information Service Providers (Jan. 1990). ERIC received an appropriation of $6.6 million in FY 1990.

[28]CRESS, Annual Review Report 14 (Jan. 31, 1991) (documents relating to migrants provided only 1.1% of CRESS input in RIE database for 1990).

[29]A recent auditor's inquiry at the Department of Education revealed that Education alone has over 700 clearinghouses of some kind (including technical assistance centers). This revelation has generated considerable interest in both the executive and legislative branches in cutting back and consolidating; creation of a new national repository would have to swim against an understandably strong tide running in the opposite direction. Interview with Patrick Hogan, Office of Migrant Education, Washington (October 22, 1991).

[30]The consultant's study to which the House Report refers itself discussed these and other centralized databases, but noted that many are badly underutilized. Adelman & Cleland, supra note 25, at 51. It also reported the comments of two state ME directors that perhaps a "saturation point" had been reached on "products" of the section 143 grant programs. Id. at 61. The study's ultimate suggestion for still another repository, id. at 62, is thus at least mildly surprising.

[31]Pub. L. No. 95-91, 91 Stat. 565 (1977).

[32]See 33 Cong. Q. Almanac 612 (1977).

[33]29 U.S.C. §§ 1800-1872 (1988).

[34]A few local service providers exemplify this approach already, at least in part. But as the description in Chapter 3 of the Iowa experience reveals, such unification still requires cumbersome multiple applications to separate funding agencies, and each of those agencies, at present, tends to insist upon clearly identifiable separate programs even when run by a common grantee. Such demands are less likely if there is complete unification at the federal level within a single department or agency. Migrant Education is harder to work into this framework, because its services are traditionally provided through public agencies, i.e., the school systems. This obstacle is not necessarily insuperable, however, for public bodies sometimes become MHS grantees or delegate agencies, and similar adaptations might conceivably allow a broader role for recipients of ME funds, if they wished to branch out.

[35]See 45 Cong. Q. Almanac 738 (1989).

[36]See 15 U.S.C. §§ 631-650 (1988).

[37]A proposal of this type appeared in a bill sponsored by Congressman Roybal in 1974, which would have created a National Office for Migrant and Seasonal Farmworkers in the Department of Health, Education and Welfare. All MSFW programs within HEW's jurisdiction would have been transferred to this office, and its work would have been supplemented by a special task force as a kind of advisory committee charged to carry out continuing studies of the needs of MSFWs and of "methods for meeting those needs." See National Office for Migrant and Seasonal Farmworkers, Hearing on H.R. 12257 Before the Subcomm. on Agricultural Labor, House Comm. on Education and Labor, 93d Cong., 2d Sess. 2 (1974).

[38]If a separate agency, it could be located within the executive branch (clearly the preferred option, given the functions it must perform) or it could be set up as an independent agency headed by a commissioner who could only be removed for cause. See generally Humphrey's Executor v. United States, 295 U.S. 602 (1935); Reginald Parker, The Removal Power of the President and Independent Administrative Agencies, 36 Ind. L.J. 63 (1960).

[39]For a stimulating argument that hierarchical reorganization is often inferior to informal coordination of "loosely coupled multiorganizational systems," see Donald Chisholm, Coordination without Hierarchy: Informal Structures in Multiorganizational Systems 1-19 (1989).

[40]Seidman & Gilmour, supra note 22, at 226.

[41] Allen Schick, The Coordination Option, *in* Federal Reorganization: What Have We Learned? at 85, 95-96 (Peter Szanton ed., 1981).

[42] Schick, id. at 86-91, identifies three main reasons why coordination is needed: planned or preferred redundancy, a pluralism of affected interests, and a lack of integrating criteria.

[43] As Schick has observed, id. at 96-97:

> The effectiveness of an interagency committee depends less on its formal status than on the extent to which member agencies share common interests and perspectives. . . . Interagency committees cannot succeed as organizational orphans. When nobody has a vested interest in the group's work and nobody is responsible for following through on its decisions, a committee will languish even if its formal status remains intact. This problem cannot be overcome merely by arming one of the group's members with "convenor" or "lead" status. The lead agency has to care enough to invest the group with resources and support.

[44] Designating positions filled by political appointees as the members of the council poses a risk of discontinuity, given the more frequent turnover in such positions. We therefore heard some suggestions for designating further members of the council drawn from the ranks of career civil service personnel. Such a move appears inadvisable. The council is likely to work more cohesively as a relatively small body. Continuity is clearly important, but if the council's functions assume any level of real importance, the policy-making officials will want to involve career officers in the body's ongoing actions, if only to prepare the member adequately for issues to be discussed at the meetings. The realities of time management also suggest that top civil servants are likely to attend some of the gatherings anyway as stand-ins for designated members.

[45] Conceivably the Council could be given its own modest staff, but such an approach risks heightening imbalances and sharpening possible resentments on the part of agencies not currently chairing the body.

[46] This is bound to be a sensitive subject, however. A council of equals, such as this option envisions, is not an auspicious forum for assuring close review of such questions; the process is potentially too threatening to all of the players involved.

[47] Such assignments of lead agencies are often provided for by statute, either designating an agency for a task directly or specifying the procedure in more general terms. See, e.g., 42 U.S.C.A. § 7521(a)(6) (West Supp. 1992) (part of the Clean Air Act Amendments of 1990,

assigning EPA lead responsibility to develop regulations, within one year, governing on-board systems for the control of vehicle refueling emissions, after mandatory consultation with the Secretary of Transportation on safety questions); 21 U.S.C. § 1504(d) (stating that the "President shall designate lead agencies with areas of principal responsibility for carrying out the National Drug Control Strategy").

[48]Pub. L. No. 100-690, tit. I, 102 Stat. 4181 (1988) (codified at 21 U.S.C. §§ 1501-1509 (1988)).

[49]44 Cong. Q. Almanac 110 (1988).

[50]21 U.S.C. § 1502(b) (1988).

[51]Id. § 1503.

[52]Id. § 1502(c).

[53]See Michael Isikoff, Martinez Suffers Setbacks as Drug Control Director, Wash. Post, Feb. 24, 1992, at A1.

[54]See id.; Michael Isikoff, Under Clinton, Drug Policy Office's Hot Streak Melts Down, Wash. Post, Feb. 10, 1993, at A14 (reporting an 84 percent staff cut, to 25 persons.

[55]Pub. L. No. 94-282, tit. II, 90 Stat. 459, 463 (1976) (codified at 42 U.S.C. §§ 6611-6618 (1988)).

[56]Oversight of the Office of Science and Technology Policy, Hearing Before the Subcomm. on Science Research and Technology of the House Comm. on Science, Space and Technology, 100th Cong., 1st Sess. 6-7 (1988) (statement of Dr. William R. Graham, OSTP Director).

[57]Id. at 7.

[58]Id. at 12.

[59]The OSTP director described this role: "In the area of budget guidance and review, OSTP interacts with the agencies of the government during the budget formation process, and then works in partnership with its fellow agency -- the Office of Management and Budget -- when the overall federal budget is prepared for review by the President." Id.

[60]Pub. L. No. 96-212, § 301, 94 Stat. 102, 109-110 (1980) (codified at 8 U.S.C. § 1525 (1988)).

[61]Id.

[62]See, e.g., Norman L. Zucker & Naomi Fink Zucker, The Guarded Gate: The Reality of American Refugee Policy 124-37; 281-82 (1987); Select Commission on Immigration and Refugee Policy, U.S. Immigration Policy and the National Interest 197 (Final Report, 1981).

[63]Interview with Dr. Luke Lee, Director of Plans and Programs, Office of the U.S. Coordinator for Refugee Affairs (October 22, 1991).

[64]Critics of placing the Refugee Coordinator's office in the State Department argue that this location risks identifying the office too much with only one of the affected agencies; the White House, in this view, is the logical site, owing to the Coordinator's transdepartmental responsibilities. The counterargument is that refugee resettlement is inevitably tied closely to overseas developments, and that the Coordinator's ambassadorial functions further support placement in the State Department. Despite the controversy, the office has remained in State throughout its existence.

[65]See Zucker & Zucker, supra note 62, at 282.

[66]Some who voiced the objection to a White House based coordinator suggested that the Coordinator's office should be independent, with the Coordinator removable by the President only for stated cause. Such an approach is not workable. Because the office would lack operational responsibilities or capacities, its initiatives must all ultimately be implemented by departmental personnel. High departmental officials have reasons to listen to initiatives that come from the Executive Office. Even if they resist, a White House-based Coordinator would be in a position (as is OSTP) to raise the dispute to a higher level for resolution. An independent body lacks this institutional clout, especially when seeking action from nonindependent executive-branch agencies. Without any hope of such leverage, a coordinator's office is essentially pointless.

[67]It is an open question whether the Constitution permits limits of this kind on the President's appointment power, particularly for a position in the Executive Office of the President.

[68]Once again, Harold Seidman has provided insightful commentary:
> The establishment of agencies within the Executive Office of the President is also sought by professions and interest groups as a means for maximizing access and influence and obtaining status and prestige. ...[But the] degree to which location within the Executive Office of the President enhances power and influence either within the executive branch or with the Congress is questionable. The power of Executive Office agencies is derived from the functions they perform -- not organization location. Influential units...are those that provide direct support to the President in conducting *presidential* business or that control action-forcing processes such as the budget and legislative clearance.

Harold Seidman, A Typology of Government, *in* Federal Reorganization, supra note 41, at 33, 39-40.

5

Data and Definitions

A new coordinating council could review program overlap, encourage enhanced cooperation at the local and state levels, work to develop consolidated outreach efforts, and explore other initiatives. But the work of all the agencies and grantees (and indeed of wholly private assistance organizations) would benefit greatly from one broader initiative – developing a better system to gather comprehensive data on America's farm worker population. Existing data use widely different methodologies and paint sharply different pictures of today's farm worker population (and their families). Many officials and service providers whom we interviewed reported unending frustration at their inability to provide legislators, budget officers, or other observers with an agreed count of migrant farm workers or of the wider category of seasonal farm workers. A central and unified system has been advocated for many years, but has been blocked by departmental rivalries, political maneuvering, and conceptual difficulties. Developing a reliable data-gathering system would help identify with assurance the real needs in the field, support advocacy for increases in overall budgets, and aid in shifting the distribution of local or regional funds available for MSFW services as farm work patterns change.

This task will require a sustained, long-term effort. Two fundamental steps must be taken at an early stage. First, there must be clarity about just who is being counted. Then, clear responsibility for ongoing data collection and analysis should be assigned to a single agency that is well-equipped to carry it out and refine the approaches over time. This chapter reviews the various definitions of MSFW now in use and the data so far employed to estimate their number and characteristics. It then sets our recommendations for developing a comprehensive system.

Myths and Stereotypes

The starting point for a discussion of definitions is a simple question: Who is a migrant farm worker? Varden Fuller, a leading student of farm labor, observed that there is a tendency "to refer to all persons who harvest crops as 'migrants' without knowing (or wanting to know) whether they were migratory or not."[1] Researchers who interview farm workers, however, describe them as "a large, highly mobile, dispersed, largely invisible population that undergoes great changes in composition."[2] Many farm worker advocates blame governmental indifference for persisting disagreements over who a MSFW is and how many MSFWs there are. These critics often note that fowl seem more important to the federal government than farm workers, since the federal government allegedly has better data on migratory birds than migratory workers.[3]

Some of this confusion is also due to the gap between the stereotype and the definition of a migrant farm worker. The stereotype is that virtually all minority workers in the fields are migrants; many definitions, on the other hand, include Iowa teenagers as migrants but not Mexican-born families settled in California who each day commute from their homes in farm worker towns to the fields. Fuller once observed that highway drivers who see a crew of Hispanic workers hoeing assume that all of the hoers are migrants, and the white tractor driver is not, while the opposite may be the case.[4]

Confusion also arises because there is a persisting myth that "millions" of people live in the southern parts of the United States and follow the ripening crops north. A typical description is that "three streams of people ... flow and fan northward, traveling from their homes around Florida, Texas, and California to distant places."[5] The map that accompanies this description has heavy black arrows which show how Florida-based migrants move up the Eastern seaboard, Texas-based migrants fan out across the midwest, and California-based migrants move within the state and north into Washington and Oregon. The arrows indicating a south to north migration of workers help to explain the nautical flavor of migrant labor discussions: states are upstream or downstream, and there are major currents and cross currents. Farm labor scholars, however, have usually emphasized that the picture of migrants flowing south to north lent a false precision to an unorganized migration and exaggerated the flow of workers. Fuller noted in 1984 that "the major change that has occurred in respect to seasonal farm labor is the decline in migratoriness ... no less important than the decline in physical magnitude is the decline in the myth."[6]

During the mid-1960s, imprecision in definitions and numbers did not seem so important because there was a sense that migrant farm workers would soon be displaced by machines. A temporary upsurge in the number of migrants had occurred in the mid-1960s, when the federal government terminated the Bracero program (see Box 1.2), but the number of migrants soon declined[7] Nevertheless, the children in migrant farm worker families were not expected to be able to follow in their parents' footsteps because of mechanization.[8] Without federal assistance, the argument ran, migrants and their children would be unprepared for nonfarm jobs. As a result, definitions of the migrant farm workers to be served, as well as the distribution of available funds, for example, between health and education services or between upstream and downstream states, were ad hoc in this era when migrancy was considered a soon-to-be-closed chapter of American history.

Migrancy, however, did not disappear. As we have seen, the number of MSFWs stabilized and even increased in some areas as labor-intensive agriculture expanded faster than mechanization displaced workers on the fewer and larger farms that accounted for most U.S. fruit and vegetable production. Furthermore, there is no indication that the number of migrant workers in the United States will decline in the 1990s; their characteristics may change, but the number of hired farm workers is likely to remain in the 2 to 3 million range, and the migrant percentage of this farm work force is likely to remain in the 30 to 40 percent range.

This fact justifies a new effort to generate reliable data on migrant farm workers. In theory, "migrant farm worker" should be easy to define. Logically, migrant is an attribute of a subset of persons who have farm worker occupations. U.S. labor force data can apply age, sex, or race attributes to workers in particular occupations, so that, in the Current Population Survey's (CPS) 1991 data on workers who have the occupation farm worker, 21 percent were women, 27 percent were Hispanic, and 9 percent were black.[9] However, the CPS does not report the number or percentage of migrant farm workers, so that the migrant attribute cannot be looked up in the same manner as age, sex, or race in regularly published labor data.

Similarly, the DOL Standard Occupational Classification (SOC) manual defines six types of farm workers, including general farm workers (SOC 5612) and both vegetable (5613) and orchard (5614) workers, but there is no mention of migrant farm workers in the SOC. The DOL Dictionary of Occupational Titles (DOT) distinguishes farm workers by the type of crop in which they work, so that workers are classified in the DOT as grain, vegetable, fruit and nut, field crop, and horticultural workers. But there are no sub-listings for migrant or seasonal workers in this industry/occupation matrix. In short, one cannot simply look up migrant farm worker in regularly published labor force data.

Bottom-Up Versus Top-Down Estimation Procedures

Since MSFW does not appear in normal labor force data sources, two major methods have emerged to adjust regularly published labor data in order to estimate their number and distribution. Most common are bottom-up estimation procedures, which begin with an enumeration or estimate of the number of MSFWs in each county or state, adjust these data to reflect MSFWs who were not included in the count or estimate, then add their dependents, and thus produce a count and distribution of MSFWs and their dependents for states and perhaps counties.[10] An alternative top-down approach begins with the total number of farm workers (or another overall indicator of farm worker activity such as wages paid to hired workers), and then adjusts these aggregate data downward to assign the subset of MSFWs of interest by states and counties.[11]

During the 1970s, most MSFW estimates were bottom-up. Studies typically began with the monthly Employment and Training Administration estimates (reported as ETA-223 data) of the number of migrant and seasonal workers employed during the week which includes the 15th of the month in areas with "significant" farm worker activity.[12] These estimates, made by local Employment Service (ES) staff, were then adjusted by the analyst to account for unemployed workers, ES undercounts of workers (as alleged by local service providers), and the dependents of the workers that the ES counted. The adjusted estimates then produced a count and distribution of the target population of MSFWs and their dependents.

Both procedures have advantages and disadvantages. Bottom-up procedures begin with data on the population of interest, but subsequent adjustments to improve these data presume that the analyst has more knowledge of MSFWs than the person who originally made the baseline estimates. Top-down procedures, by contrast, usually begin with more reliable data, but they must make often arbitrary assumptions in order to isolate the target subset of farm workers.

The problems inherent in both bottom-up and top-down procedures have prevented either from emerging as the generally accepted method for studying farm workers. Even worse, from an analytical perspective, few studies using either procedure have ever been done twice (the usual practice for cross checking on the accuracy and reliability of the method) so that, in the case of MH, studies done in 1973, 1978, 1985, and 1988 were in no way cumulative or self-correcting.[13] It appears that several hundred thousand dollars have been spent annually by federal MSFW assistance programs to estimate the number and distribution of MSFWs and their dependents, or perhaps $5 million since the mid-1970s, but there is still no agreement on the number and distribution of MSFWs.[14]

The experience with these studies has not even produced agreement on a procedure to determine their number and distribution.

There has been relatively little progress made to develop reliable data on MSFWs, and there is no agreement on whether better data could be obtained with bottom-up procedures, top-down procedures, or a combination. There are several options. An improved bottom-up procedure might, for example, build on an improved MSRTS (Migrant Student Record Transfer Service, described in Chapter 2). An improved top-down procedure might be based on a refined Census of Agriculture (COA), a modified decennial Census of Population (COP), a revised Current Population Survey (CPS), or an expanded National Agricultural Workers Survey (NAWS).

Program service data today do not reflect the total migrant population because each program serves a unique subset of farm workers, according to its own specific definitions and eligiblity standards, and no program serves all of the eligible MSFWs and dependents. Each program's target population is like a circle, and the service data refer to the quarter or half of the circle that program staff see. The target population circles of each program partially overlap, but no one has been able to determine whether the partial circles of persons served are the same people or not. A few service providers that operate several of the Big 4 programs asserted that, in their experience, as many as 80 percent of those considered migrant workers by one program are also considered migrant workers in other programs, but there is no easy way to check how widespread this eligible-for-one and eligible-for-all experience is.

A uniform federal definition of MSFW, applicable to all the programs, could make administrative data on who was served a useful element in an improved bottom-up estimation procedure, but even service data from a uniform definition cannot fully establish the number and distribution of migrants. Estimates of the total migrant population based on persons served may miss migrants in areas not currently touched by assistance programs and may simply reflect levels of outreach and funding. For these reasons, existing bottom-up procedures are questionable, and they leave problems that even a uniform definition cannot cure, at least without resource commitments assuring near-total coverage of the target population. Therefore, it is preferable to determine the number and distribution of MSFWs from a census or sample survey -- a top-down procedure -- rather than from the enrollment data of programs.

Three Suggestions for Change

Three proposals have been made to generate better data on MSFWs. First, some of the JTPA 402 program participants, whose funding is based on Census of Population (COP) data, believe that a slight modification of 1 or 2 COP questions could make the COP a valuable source of data on farm workers. (The decennial Census of Population is not currently considered a reliable vehicle to estimate the number and distribution of especially migrant farm workers because the COP asks respondents about the work they did in the week before the Census, and the last week in March finds employed only one-third of the people who do farm work during a typical year.[15]) JTPA grantees propose that the COP ask respondents what amount or percentage of their earnings in the year preceding the COP were from farm work. In this way, a respondent not employed as a farm worker in March but with farm earnings during the previous 12 months, can be identified on the COP.

This change in the Census questionnaire would provide useful data on farm workers, but it will not produce the number and distribution of MSFW data that are needed. For example, knowing that a worker had farm earnings during the previous year does not distinguish migrant workers from other farm workers. Moreover, assistance programs need to track changes in farm worker population much more closely than a decennial snapshot permits.

Second, better data on the number and distribution of migrant farm workers could be obtained by combining several sources of data in a top-down procedure. One top-down study combined state-by-state data from the Census of Agriculture (COA), regional data from the Quarterly Agricultural Labor Survey (QALS), and data from the Current Population Survey (CPS) on worker characteristics to distribute farm worker activity across states. This study suggested that there may be 600,000 to 1.2 million migrant farm workers in the United States, depending on the exact definition of who a migrant is.[16]

Since this top-down distribution of farm worker activity was completed in the mid-1980s, a new national survey of farm workers has been undertaken. The National Agricultural Workers Survey (NAWS) is a national worker survey established by the U.S. Department of Labor after the enactment of the Immigration Reform and Control Act of 1986

Figure 5.1. NAWS Profile of About 2 Million SAS Farm Workers: 1989-1991[17]

1. Demographic Characteristics: Most farm workers are male, young, married, and immigrants with a SAW status

   - 73 percent are male; 67 percent are 35 or younger (median age 31); but 17 percent are 20 or younger
   - 58 percent are married; 52 percent have children, but 40 percent do farm work unaccompanied by their families
   - 60 percent are foreign born, including 55 percent who were born in Mexico
   - 70 percent are Hispanic
   - 29 percent are SAWs (580,000 of an estimated 2 million), and they have a median 7 years of U.S. farm work experience
   - 10 percent are unauthorized (200,000 of 2 million) and these young workers (median age 23) have only 2 years of U.S. farm work experience
   - 53 percent of all SAS workers have 8 or fewer years of education (median 8 years); 65 percent speak primarily Spanish
   - 57 percent of the workers live with their families at the work site; 85 percent of the persons 15 and older in households work

2. Farm Work and Earnings: Most workers experience extensive seasonal unemployment and have low annual earnings

   - Workers have a median 8 years experience doing SAS farm work; U.S. citizens and greencard immigrants average 11 to 12 years, while unauthorized workers average just 2 years experience
   - Workers on average spend 50 percent of the year or 26 weeks doing Seasonal Agricultural Services work for 1.7 farm employers, 20 percent or 10 weeks unemployed, and 15 percent or 8 weeks doing non-SAS work and 15 percent or 8 weeks traveling abroad
   - 75 percent of all crop workers are employed in fruits or vegetables
   - 77 percent are hired directly by growers, usually to harvest crops
   - Median earnings were $4.85; work weeks averaged 37 hours, for SAS earnings of $180 and, for 26 weeks, $4,665
   - Less than half of the workers have Unemployment Insurance and Workers Compensation coverage; 21 percent have off-the-job health insurance
   - 28 percent live in employer-provided housing

3. Other Work and Income: Farm workers are poor but not dependent on welfare

   - 46 percent of all SAS workers have below poverty level incomes, the poverty rate is highest for unauthorized workers (77 percent)
   - 36 percent also do non-SAS farm work; such work paid a median $4.50 per hour and is preferred to farm work
   - 58 percent are unemployed sometime during the year; 50 percent of these unemployed are jobless less than two months; only 28 percent of the jobless farm workers apply for unemployment insurance
   - 40 percent of the workers spend an average 19 weeks abroad each year
   - Median individual incomes are $5,000 to $7,500; median family incomes are $7,500 to $10,000; 50 percent of the families are below the 1989 poverty line of $12,675 for a family of four
   - 55 percent of the workers own assets, usually a vehicle
   - 16 percent get food stamps; 3 percent Aid to Families with Dependent Children

Source: Richard Mines, et. al., Findings from the NAWS 1990 (Washington: U.S. DOL, Office of Program Economics), Research Report Number 1, 1991 and additional data analysis. These data are preliminary, and their interpretation is our own.

(IRCA)[18] to help to determine whether there were farm labor shortages that required the admission of additional agricultural workers. The NAWS has a top-down sampling strategy in the sense that it relied on COA data to determine the counties where farm workers should be interviewed, and then used QALS data, supplemented with additional lists of farm employers, to determine where to interview farm workers.[19] The NAWS then interviewed workers who were employed, and obtained work histories from them.

NAWS data could be used to determine the number, characteristics, and distribution of migrant, seasonal, or any other farm worker subgroup of interest. For example, if migrant workers are defined as those who travel at least 75 miles from their usual residence to do farm work, then 42 percent of the workers interviewed between 1989 and 1991 were migrants. Most of these migrant farm workers were immigrants who shuttle into the United States from homes in Mexico.

The NAWS has no sampling frame in the sense that it did not begin with a list of the nation's 2 or 3 million farm worker households and interview 1 percent of them. For this reason, NAWS data cannot be readily expanded to assert, for example, that there are 840,000 migrant farm workers in the United States (42 percent of the estimated total 2 million crop workers). However, an expanded NAWS – a survey that interviewed 20,000 or perhaps 1 percent of all workers instead of the 7,000 actually interviewed – might provide a national source of data on farm workers from which the migrant subset could easily be validly determined.

A third suggestion, and an alternative to an expanded NAWS, is to upgrade the monthly Current Population Survey (CPS). CPS interviewers visit 60,000 U.S. households to determine who lives in the household, whether household members work and, if they do, their industry and occupation of employment, hours of work, and earnings. These CPS data are the basis for national and state unemployment rates.

Until 1987, the CPS included supplemental questions in December which asked if anyone in the household had done farm work for wages during the calendar year. About 1,500 households in the December CPS included a farm worker. For these farm worker households, data were collected on where the farm worker worked during the year as well as his farm and nonfarm earnings. The U.S. Department of Agriculture (USDA) analyzed these CPS data to estimate the number and characteristics of migrant farm workers.

The USDA defined migrants as persons who crossed county or state lines and stayed away from home at least one night during the year to do farm work for wages. USDA defined farm work to include crop and livestock agriculture, but to exclude the processing of crops and livestock, and USDA imposed no occupational, earnings, or legal status criteria on who could be a farm worker. As a result, veterinarians as well as field hands could be migrants, and legally authorized as well as illegal alien workers were included in the migrant count. Teenagers in Hispanic families who migrated from Texas to Michigan could be migrants, as well as Iowa teenagers who lived and worked on an uncle's farm in another county during the summer. According to USDA's analysis of CPS data, there were 115,000 to 226,000 migrant farm workers in the United States in the 1980s; most of them were in the midwestern and southeastern states.[20]

Many knowledgeable observers found these USDA reports suspect. Farm worker service providers and researchers complained that the data showed too few migrant workers, and far too few Hispanic migrant workers. These critics argued that the design of the CPS prevents it from accurately counting migrant workers. The CPS is constructed so that each of the nation's roughly 92 million housing units has an equal probability of being interviewed, so that each farm worker interviewed can be assumed to represent another 1,300 or so who were not interviewed. But critics of the CPS argued that especially Hispanic migrant workers tend to live in the kind of nontraditional housing that CPS lists and interviewers are likely to miss, especially in December when many Mexican migrants have returned to Mexico.

Just as the NAWS could be expanded to generate a more reliable profile of migrant workers, so the CPS could be modified to generate better data on migrant workers. For example, if the major problem with the CPS is that supplemental questions about farm work were asked in December, when many migrants are not in the United States, then the supplemental questions could be asked in September, the month of peak farm worker employment. If the problem with the CPS is that its interviewers are not sufficiently sensitive to the language and culture of migrant farm workers, appropriate interviewers can be hired or trained.

Migrant Numbers and Characteristics

What do the COP, NAWS, and CPS say about migrant farm workers? The 1990 COP data on farm workers are not yet available for all states, but the 1980 COP reported that there were 875,000 farm workers (Occupation 479). Two-thirds of these farm workers were White and, of the 307,000 minority farm workers, one-third were Black and two-thirds were Hispanic. Almost 60 percent of the 200,000 Hispanic farm workers were in California.

Similar data from the 1990 COP are available for some of the larger farm states. In California, for example, the 1990 COP reported that there were 177,000 farm workers; in Texas, 68,000, and in Florida, 44,000.

However, the COP April snapshot may miss up to 75 percent of the farm workers in these states, and the workers most likely to be missed in the COP are minority migrants.[21] Modifying the Census questionnaire to ask respondents to report whether they had any farm earnings throughout the year would presumably increase the number of migrant and seasonal workers who are counted in the Census, but the COP may still miss minority farm workers in the same way that it reportedly misses other ethnic minorities.

NAWS Data

The NAWS profiles a different farm work force. Instead of finding mostly white farm workers, the NAWS found that three-fourths of all farm workers in Seasonal Agricultural Services (SAS) or U.S. crop agriculture[22] are minorities, usually Mexican nationals who have been in the United States for less than 10 years (Figure 5.1). The NAWS found that most of these farm workers are married men who are poorly educated and who live with their families at their U.S. work sites.

In surveys conducted between 1989 and 1991, the NAWS found that 40 percent of SAS farm workers were U.S. citizens and 60 percent were aliens. The aliens included Special Agricultural Workers (SAWs)-- 40 percent had "temporary resident" status; another 25 percent were (greencard) permanent resident aliens; 16 percent received legal status through other programs, such as political asylum; and about 20 percent of the alien workers were undocumented. About 60 percent of the estimated 200,000 unauthorized alien workers arrived within the past five years, i.e., after IRCA was enacted.[23]

The average Seasonal Agricultural Services (SAS) worker did farm work for 1.7 farm employers. Most workers were employed for about half of the year at hourly wages that were 14 percent above the federal minimum of $4.25 hourly. NAWS workers averaged $4.85 for 37 hours of work per week while they were doing farm work. Their average weekly earnings of $180 for 26 weeks of SAS work generated annual farm earnings of $4,665, about three-fourths of the $6,465 poverty-level income for an individual in 1990. SAS farm workers average another 10 weeks of unemployment searching for farm and nonfarm jobs.

The NAWS found that most Seasonal Agricultural Services (SAS) workers were employed in fruit, vegetable and horticultural commodities (FVH); 90 percent worked in these commodities. Vegetable farms employed 43 percent of the sample workers sometime during the year; fruit and nut farms 32 percent; and nursery and horticultural specialty operations 15 percent. This distribution of workers across commodities differs significantly from the distribution of labor expenditures in the Census of Agriculture. In the COA, almost 40 percent of all crop labor expenditures were made by non-FVH farms and, in a further example, vegetable farms accounted for just 24 percent of COA labor expenditures, while the NAWS reported that 43 percent of the workers interviewed were employed by vegetable farms. [24]

Over one-third of the farm workers in the NAWS also did nonfarm work. Even though it paid slightly less per hour than farm work, nonfarm work was preferred by most workers interviewed while doing farm work, suggesting that layoffs or a lack of nonfarm jobs pushed these workers into farm work, rather than the attraction of higher wages. Nonfarm work in services such as janitorial or clean-up businesses or in construction was reported by many farm workers to be more stable and to offer them more opportunities for upward mobility.

Many farm workers spend part of each year abroad; the average SAS worker spent 8 weeks abroad during the year. It has been argued that most workers from Mexico are small farmers who come north when drought or pests wipe out their crops. NAWS data seem to belie this notion that Mexican farm workers in the United States use the U.S. farm labor market as a safety valve; according to the NAWS, even the workers who returned to Mexico for longer periods did not earn much money there.[25] Instead, most of the time abroad is devoted to vacation and visiting relatives.

Aliens were 60 percent of the crop workers in the early 1990s, and an even larger share of the new entrants into the work force. By some estimates, 90 percent or more of the persons who have been doing farm work for less than 3 years were born abroad, and many of the U.S. citizens among those who have just entered the farm work force have alien farm worker parents.

The NAWS was designed to be a national sample. For this reason, there are no NAWS data for states or regions. However, 25 percent of the interviews were conducted in California, and a preliminary tabulation of the California NAWS data[26] suggests that the state's farm workers included in the NAWS are similar in age and marital status to farm workers throughout the United States. California has a higher percentage of immigrant workers (90 versus 60 percent), a higher percentage of SAW workers (62 versus 29 percent), and more poorly-educated workers (71 percent with 8 or fewer years, versus 53 percent). As in the national NAWS data, California crop workers are mostly young men who are married and have children, but about one-third of them do not have their children with them when they are doing farm work in the state.

California workers did one-fourth more weeks of SAS work than did similar workers throughout the United States. While doing such work, they had 12 percent higher hourly earnings (median $5.41 versus $4.85) but, because they worked more hours per week (41 versus 37) and more weeks per year (33 versus 26), they had 57 percent higher SAS earnings ($7,320 versus $4,665). An astounding 6 of 7 worked sometime during the year in fruits and vegetables (this may also include nurseries, but it excludes livestock and dairy). Most California farm workers believed that they were covered by workers compensation (WC), unemployment insurance (UI), and health insurance, but since under California law virtually all farm workers should be covered by WC and UI, it is surprising that one-third or more reported that they were not covered.

Almost half of the NAWS workers interviewed in California had incomes below the poverty level, about the same percentage as farm workers throughout the United States. Fewer California workers have nonfarm work throughout the year (17 versus 36 percent) and, as elsewhere in the United States, this nonfarm work offered lower hourly earnings than farm work. In California, median nonfarm hourly earnings of $4.92 were 10 percent less than farm earnings of $5.41. California farm workers were about as likely to spend time abroad as are farm workers throughout the United States; the half who spent time abroad spent an average 4 months abroad.

NAWS data are probably the most reliable demographic data available on 80 percent or 2 million of the nation's estimated 2.5 million hired farm workers, but they have several limitations. First, the NAWS can be reported only as percentages within the sample. This means that the NAWS finding that 92 percent of the crop workers interviewed in California are Hispanic aliens cannot be translated into the statewide total of foreign-born and U.S.-born workers because the NAWS does not know how many farm workers there are in the state.

Second, the purpose of the NAWS was to estimate days of farm work gained and lost due to entrants and exits from the crop or SAS work force, and then only at national levels. Although 25 percent of the interviews are conducted in California, the survey was not designed to produce state or regional farm worker data, and none have been published. Third, there are no public use NAWS data available. What is known about farm workers from the NAWS is what the U.S. Department of Labor has made available. There is little opportunity to organize NAWS data differently, and thus it is hard at this time to compare available NAWS data to CPS, UI, and other data.

143

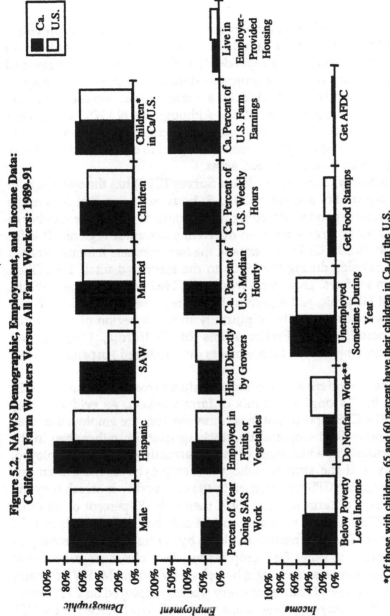

Figure 5.2. NAWS Demographic, Employment, and Income Data: California Farm Workers Versus All Farm Workers: 1989-91

*Of those with children, 65 and 60 percent have their children in Ca./in the U.S.
**Nonfarm work includes livestock work. Source: Interviews with 1,844 SAS or crop workers in California and 7,242 workers throughout the United States between October 1989 and October 1991. These data are preliminary, and their interpretation is not necessarily that of the U.S. Department of Labor staff who are collecting and analyzing these data.

CPS Data

The NAWS profiles a farm worker force that most service providers are familiar with, but its sample results cannot be extrapolated to distribute farm workers and dependents eligible for services to states and counties, as many assistance programs desire. Furthermore, it may be difficult to modify the NAWS in a manner that would allow the generation of such geographic data. Although the NAWS has emerged as the major source of information on the demographic characteristics of farm workers, it has not yet proven to be useful to allocate limited MSFW assistance funds to state and local areas.

The monthly Current Population Survey (CPS) has the advantage of being a probability sample of all U.S. households, so that its sample results can be expanded according to a formula to report the number of various types of farm workers in different geographic regions. It is for this reason that the CPS could expand the farm workers it found in about 1,500 households during the 1980s to the estimated total 2.5 million hired farm workers. These farm workers in 1987 were 78 percent White, 14 percent Hispanic, and 8 percent Black and other (Figure 5.3). White workers were the majority or plurality in every region of the United States, including the Pacific states of California, Oregon, and Washington, where white farm workers outnumbered Hispanics 54 to 44 percent.

Service providers and researchers allege frequently that the CPS generates the "wrong" ethnic mix of farm workers. As evidence, they note that the CPS reports that most farm workers are employed in non-labor intensive field crop and livestock agriculture, rather than in the FVH agriculture in which most of those currently served are employed. For example, if the workers who were employed in more than one commodity in the CPS are assigned to the commodity in which they did the most days of farm work in 1987, then only 20 percent of all hired workers – 518,000 of 2.5 million – worked only or mostly or only in FVH commodities. Over 1.1 million workers, by contrast, worked mostly or only on grain and other field crop farms. Few observers believe that there are two workers who might be helped by federal MSFW programs employed on non-FVH farms for every worker employed on FVH farms.

The CPS portrays a young, white, and male work force. In the CPS data, whites are the youngest farm workers (median age 24), largely because so many of them are 14 to 17 year-old teenagers in the midwestern states. However, in the regions such as the Pacific and the southeastern states that produce FVH commodities, farm workers outnumber operators, and the operators tend to be white while the hired workers tend to be minorities.

Figure 5.3. CPS Profile of 2.5 Million Farm Workers: 1987

1. Demographic Characteristics: Most farm workers are young White men.
   - 80 percent are male; 70 percent are 34 or younger (median age 26); and 2 percent are 14 to 17 years of age.
   - 78 percent are White; 14 percent are Hispanic; and 8 percent are Black and other; Whites are the youngest farm workers (median age 24), Hispanics (29), and Blacks oldest (30).
   - 23 percent of all hired farm workers have 8 or fewer years of schooling; 50 percent had 12 or more years of education. Hispanic farm workers had the least education: 56 percent have 8 or fewer years of education, versus 35 percent of Blacks and 16 percent of Whites. Among farm workers 25 and older, 26 percent had 8 or fewer years of education, including 72 percent of the older Hispanic workers.
   - Midwestern Corn Belt states such as Iowa and Illinois include 19 percent of all farm workers, followed by the Pacific states of California, Oregon, and Washington, and then the Lake states of Wisconsin, Minnesota and Michigan (12 percent each).

2. Farm work and Earnings:
   - Workers averaged $3,400 for 112 days of farm work in 1987, and $3,300 from doing nonfarm work, for $6,700 total earnings. Daily farm earnings of $30 suggest an average wage of $3.76 hourly for an 8 hour day.
   - Hispanics had the highest average daily farm earnings ($32) and did the most days of farm work (139), so their $4,500 annual farm earnings exceeded those of Whites ($3,200) and Blacks ($2,700).
   - Most farm workers worked primarily in grains and field crops (45 percent), or with dairy or livestock production (34 percent). Only 21 percent worked mostly with FVH commodities. Almost 50 percent of the Hispanic farm workers worked with FVH commodities, and almost one-third of the Black farm workers worked in tobacco and cotton production.

Source: Victor J. Oliveira and E.J. Cox, The Agricultural Work Force of 1987: A Statistical Profile (USDA Economic Research Service, 1990).

A Serious Strategy for Farm Worker Data

For the COP, the NAWS and the CPS to generate such sharply different pictures of the U.S. farm worker population is discouraging – to demographers, but also to policy makers and service providers. If there is little agreement on the scope of the problem that policy and assistance programs are to address, it is far harder to make headway toward their overarching goals, or indeed toward refining and revising those goals in light of what is really happening in America's fields.

These inadequacies have been bemoaned and condemned for years. It is time for a serious medium-term strategy to assure the development of an adequate database covering America's farm workers, including at long last the ability to identify and track that subset of farm workers and their dependents who are eligible for federal assistance programs. (This means primarily, but not exclusively, identifying those who are migrants or who recently migrated.) The benefits to assistance programs, and also to legislators and administrators, would be substantial.

Two basic steps are needed. First, a uniform core definition of farm worker should be developed, and similar uniformity should be established to identify migrant farm workers and other subgroups of principal interest to assistance programs. (We propose such uniformity primarily for centralized data-gathering purposes, but it could well happen that such a focus on core definitional elements might help promote, over the long run, a convergence of eligibility standards for the various assistance programs as well.) Second, a comprehensive data-gathering infrastructure should be established, replacing the present system that scatters responsibility. We recommend that the assignment be given to the Bureau of Labor Statistics, the same agency that for decades has generated nonagricultural employment data, such as the monthly unemployment figures. It is time to end farm workers' exclusion from these important statistical systems.

Developing a Uniform Definition

As we have seen, there is no generally accepted federal definition of a farm worker, and each of the programs that provide services to farm workers and their families has a different definition of the term "migrant farm worker." To help move toward uniformity, we here review the concept of the industry agriculture, the occupation farm worker, and the subset of farm workers who are considered to be migrant or seasonal workers. Where choices must be made among competing definitions, we have used a general rule of thumb, derived from the basic purpose of this effort: *viz.*, that data are needed primarily to facilitate the work of farm worker assistance programs.

One of the key factors justifying such specialized programs for farm workers is their exclusion from certain federal protective labor laws, especially those governing collective bargaining and overtime wage protections. This "farm worker exceptionalism," whatever its merit when these programs were enacted over fifty years ago, survives now more as a result of the political realities in American agriculture than as a sensible way to regulate a portion of the labor market. But as long as it persists, it remains a central element in the justification for special farm worker assistance programs. Hence our definitional elements generally seek to track the contours of those exceptions.

The Industry Agriculture

Agriculture is the production process that occurs on farms, just as manufacturing is the production process that occurs in factories. However, not all agricultural production processes take place on easily recognizable farms. Most farms have barns, fences, and other familiar features, but many hired workers are employed on "farms" that have their headquarters in urban office buildings. There are often no familiar farm buildings in the orchards or vineyards where many migrants workers are employed, and no buildings at all in the forests where other migrant workers find jobs.

Farm employers are sometimes just as hard to recognize as farms. Most agricultural data assume that there is one operator per farm, and that anyone else employed on the farm is an unpaid family worker or a hired worker. While this 3-way classification describes accurately the employment situation on most U.S. farms, it is increasingly being outmoded by the growing tendency of farms to rely on agricultural service firms to supply labor and other inputs they need. For example, a worker earning wages for picking peaches can be an unpaid family member, a worker hired by the farm operator, a worker brought to the farm by a Farm Labor Contractor (FLC), or an employee of a peach packing shed or some other nonfarm enterprise. Even though the same work is being done, differences in who is considered the employer determine whether the worker is considered a farm worker and counted in the various data sources.

The U.S. economy is divided into different industries by the Standard Industrial Classification Manual (OMB, 1987). The SIC divides operations that engage in agricultural production into five categories: crop production (SIC 01), livestock production (02), agricultural services (07), forestry (08), and fishing, hunting, and trapping (09). Establishments are classified into one of these categories on the basis of their sales, so that a combination strawberry and Christmas tree farm, if at least 50 percent of its sales are from strawberries, is classified first as a crop farm (01), then as a fruit and nut operation (017), and finally as a berry establishment (0171). In the economic censuses conducted every five years, any employment-related data provided by establishments to government agencies are included in only one SIC code, so that any Christmas tree workers and wages from the combination tree and berry farm described above are reported with strawberry crops.[27]

The SIC groups farm operators in terms of the products they sell. If a worker is employed directly by a peach farmer, he is a crop worker. However, if the worker is employed to pick peaches by a FLC, he is an agricultural services worker and, if employed by a peach packing shed, perhaps a nonfarm worker.

Data reported by the type of farm operation define workers differently than do labor laws, which determine who is covered and not covered by the type of work activity done and where it is performed. This distinction is apparent in the Fair Labor Standards Act (FLSA), the major federal law regulating minimum wages, overtime, and child labor. The FLSA excludes farm workers from its overtime protections with a primary and a secondary definition of agriculture:[28]

"'Agriculture'" includes [*primary definition*] farming in all its branches and among other things, includes the cultivation and tillage of the soil, dairying, the production, cultivation, growing, and harvesting of any agricultural or horticultural commodities (including commodities defined as agricultural commodities in section 1141j(g) of Title 12), the raising of livestock, bees, fur-bearing animals, or poultry, and [*secondary definition*] any practices (including any forestry or lumbering operations) performed by a farmer or on a farm as an incident to or in conjunction with such farming operations, including preparation for market, delivery to storage or to market or to carriers for transportation to market.[29]

These definitions emphasize that agriculture includes both the production and the processing activities that occur on a farm, and that workers must be employed on a farm to be considered for farm worker status.

The National Labor Relations Act (NLRA) also excludes agricultural laborers. Although the NLRA does not define the term agricultural laborers, Congress has specified in appropriations legislation that agricultural laborer is to be understood as it is defined in the FLSA.[30]

One way to define the core population of concern to MSFW assistance programs would be to focus on those who are employed (or have parents who are employed) as farm workers in agriculture as defined by the FLSA or NLRA. Such a definition has an internal logic – the federal government's exclusion of farm workers under these labor laws makes it harder for them to help themselves, so that MSFW assistance programs are partial compensation for their exclusion from labor laws.

Many assistance programs go beyond farm worker as defined by federal labor laws to make eligible those workers and their dependents who are employed by nonfarm food processors and packers, wholesale food handlers, and some combination growing and retailing operations.

This dramatically expands the pool of potentially eligible workers. The U.S. food manufacturing industry (SIC 20) employed an average 1.6 million persons in 1988, including 250,000 in canned and frozen foods (mostly fruits and vegetables). Since there is high turnover in many food manufacturing establishments such as meat and poultry processing, the number of persons employed sometime during the year is considerably higher, perhaps 2 to 3 times average employment. Another 822,000 were employed in wholesaling food items, including firms, for example, that buy tomatoes from farmers and repack them.[31] Since employment in farm-related industries exceeds employment on farms, it is clear that making eligible for services workers in nonfarm industries related to farming could lead to a situation in which the majority of participants in a migrant assistance program were "nonfarm" workers under federal labor laws.

The lack of a common definition of agriculture or qualifying work is a fundamental obstacle to developing better data on farm workers in the usual building-block fashion of doing repeated surveys and accumulating a better understanding of the target population. It also hinders the development of yardsticks against which the problems and progress of the farm workers eligible for services can be assessed.

The Occupation Farm Worker

Once agriculture or qualifying employment is defined, intake workers employed by the various assistance programs must determine who, among all employees, is eligible for assistance. For example, eligible farm workers can be defined as all persons who do qualifying work for wages, or eligibility can be restricted to employees who are not related to the farm operator.[32] Farm workers are often defined by the industry in which they work--as when all persons employed on farms are "farm workers" for FLSA purposes. Farm workers also be defined by the task they perform, so that on a large FVH farm, there may be fieldworkers and equipment operators, supervisors, as well as accountants, lawyers, pesticide advisors, and other professionals. The number of such workers who are employed on farms but who perform nonfarm tasks is considerable; in California UI data, about 20 percent of the weeks of UI benefits paid by agricultural employers (SIC 01, 02, 07) are paid to persons who do *not* have farm worker occupations, such as mechanics or computer operators employed on large farms.[33]

Agricultural operations have become far more complex since the mid-1960s, when "eligible farm worker" typically meant everyone employed on farms, but there is no easy way in regularly published labor data to distinguish field workers from other employees on farms. The nation's major effort to collect data on who is employed is the federal-state Current Employment Statistics (CES) program supervised by the U.S. Department of Labor's Bureau of Labor Statistics. Data are collected monthly from 350,000 establishments that employ 41 million workers, and month-to-month changes in payroll employment are announced by BLS with the unemployment rate.

The BLS excludes most farm workers from the CES. Because its mission focuses on nonagricultural employment, farm operators are not in the BLS sample of reporting establishments. There is thus no way to get BLS to revise an ongoing data series in order to determine the number of farm workers, as opposed to the number of nonfarm workers, on the payrolls of agricultural establishments. However, such a reporting system can be developed. California has developed an agricultural CES system, for example, that collects monthly data on employment of wage and salary and production workers from approximately 12 percent of the state's 25,000 farming operations and farm service businesses that pay UI taxes on the wages of hired workers. In 1992, there were an average 348,000 wage and salary workers employed by such agricultural employers, including 320,000 production workers. This California experience demonstrates that BLS could add agricultural operations to its employment-reporting system and distinguish subgroups of farm workers.

Having BLS estimate the number of farm workers employed each month may be only a first step toward isolating the subgroup of workers who may eventually be deemed eligible for federal assistance. For example, most service programs require a qualifying move or period of employment that can be as short as one day, while others require that to be eligible for services, farm workers must have been employed in agriculture for at least a minimum period of time -- such as 75 days -- and/or to have earned at least half of their previous 12 months income by doing farm work. It would not be possible to isolate such subsets of farm workers in BLS data. Consequently, even if BLS were to begin collecting data on farm workers, there would still be a need for intake workers in the specific assistance program to determine whether a worker satisfies eligibility criteria in addition to being a farm worker.

The Elements of a Core Definition

The previous sections demonstrate the wide array of options available in determining just who is to count as a farm worker and ultimately in deciding who should be eligible for the special assistance programs designed for MSFWs. For data to be useful and cumulative over time, there must be agreement on central elements of the definition of the people who are being counted. Of course, there need not necessarily be a single identified category (for data purposes at least); one could generate separate counts of crop vs. livestock workers, for example, or of those involved in field work and on-farm processing vs. other food processing workers. Nevertheless, we offer here our views on how best to select core characteristics for profiling and tracking the most relevant subgroups of farm workers. (See Table 5.1).

Four basic steps are required. First, to decide on *qualifying employment*, one of the available definitions of agriculture must be employed. As we have seen, agriculture can be defined narrowly or broadly. Traditionally, Migrant Head Start has used one of the narrowest definitions, covering only the production and harvesting of tree and field crops – thus excluding, for example, livestock and dairy workers. Migrant Education, by contrast, has one of the broadest definitions, embracing both crop and livestock agriculture, dairy, fisheries, and potentially must of the food processing industry.

We recommend that qualifying employment be limited to the industry agriculture as defined by the FLSA, precisely because that definition works to exclude farm workers (as there defined) from many federal labor law protections. Those agricultural workers who do not enjoy federal rights to collective bargaining, overtime wage, and similar protections, may be less able to help themselves; hence they are most likely to have to call on the aid available through MSFW assistance programs. Core farm worker data systems should focus on this more needy group.

If this standard were also one day adopted for eligibility purposes by all the assistance programs, it would modify the standards of virtually all of them. More workers would be served by programs such as MH and MHS, since they currently exclude the livestock workers who are included in the FLSA definition of agriculture. Fewer workers would be served by ME and JTPA 402 programs; these programs would have to stop serving workers and their families employed in nonfarm packing and processing establishments. The net effect of these changes in qualifying employment would be to reduce the number of persons eligible for services, since far more are employed in nonfarm-related establishments than are employed in livestock agriculture.

The second definitional issue requires a decision regarding *which workers* in agriculture, as the latter term is defined, will be counted. The question really is whether all persons employed for wages in agriculture at any point during the stated census period will be considered, or whether a narrower group is more appropriate. We recommend that the core definition count only those who have demonstrated a commitment to farm work – by doing such work for at least half of his or her working time, or by generating half of his or her earnings through qualifying farm work.

Our recommendation comes closest to the current eligibility criteria of the JTPA 402 program; its eligibility requirements include an individual work history that indicates that at least 50 percent of the work days or earnings during 12 of the previous 24 months were in or from farm work. We find these windows within lookback periods overly cumbersome, and we recommend that an individual seeking assistance services satisfy the 50 percent rule for the previous 24 months.

Adopting our recommendation that an eligible farm worker is a person who satisfied either the 50 percent of work days or earnings from qualifying employment in agriculture during the previous 24 months would widen and narrow current definitions. JTPA 402 programs would be least affected. It would be easier to qualify for MHS services than at present, but harder to qualify for ME and possibly MH services. These changes should not place an undue administrative burden on assistance programs, since most already collect work histories.

Our recommendation of eligible farm worker does not disqualify for services long-season and year-round workers. Although some assistance programs attempt to restrict eligibility to those employed less-than-year-round, we are not persuaded that an effort to distinguish long-season from year-round farm workers in a definition is worthwhile.

Within the category of eligible farm worker, most assistance programs give priority to migrant workers. It therefore makes sense for the new data system to develop separate information on *migrants*, i.e., those who established a temporary home in order to do farm work. However, there is a variety of opinion on what type of move is disruptive enough to give workers and their dependents priority for assistance services. We recommend that a migrant be defined as a person who: (1) moved at least 75 miles from a usual residence and (2) stayed away from home overnight in order to accept qualifying agricultural employment. This recommendation is close to the USDA migrant definition: we substituted the 75-mile criterion for the USDA cross-a-county-line rule to avoid creating artificial migrants in border areas,[34] and because census interviewers who are recording work histories can easily determine whether the worker moved the qualifying distance.

Accepting our migrant farm worker definition for eligibility purposes would generally narrow the definition of who gets priority for assistance services. Most of the assistance programs are vague about migrancy, and none include both the elements of distance moved and an overnight stay.

The fourth element of the core definition is to establish the *duration* of eligibility, or to determine how long after a migratory event or the necessary farm work is done workers and their families should still be counted as migrant workers or as farm workers. Although this is primarily a question of eligiblity determinations, we recommend a 24-month lookback period, which means that the persons are still counted as migrants or farm workers up to 24 months after they migrated or did qualifying farm work.

There is no way to establish definitively how long after migration or farm work lives of workers and their families are disrupted. Most assistance programs initially restricted services to 12 months after the last qualifying employment, under the theory that priority should be accorded to workers in-the-stream. But service providers noted that it was easier to help migrant workers and their families after they had settled out, and that it made little sense to push people out of assistance programs just when the assistance was most likely to be effective.

We are mindful of this tension between helping those who are currently moving and providing assistance when it can do the most good to change peoples' lives. We believe a 24 month duration of eligibility is the best compromise between these competing goals. Since it is also the most common duration-of-eligibility criterion used by assistance programs, it also requires the fewest changes.

A core definition of MSFWs and their dependents is not a cure for ad hoc growth or inadequate coordination between farm worker assistance programs. But in recognition of the changes that have occurred in both the nature of farm jobs and the characteristics of farm workers, our recommended core definition should help to put these programs on a firmer footing for the 1990s. Migrant farm workers remain needy, and farm workers are still excluded from federal labor law protections, and until the federal government can remove some of the legal obstacles that hinder farm workers from helping themselves, specially focused assistance programs are justified. But these assistance programs must adapt to the changing industry which employs their immigrant clients.

Table 5.1. Recommended Core Definition of MSFW

Item	Recommendation	Justification	Effect
1. Agriculture/ qualifying employment	Employment in agriculture as defined in FLSA.	The FLSA excludes and provides fewer protections for farm workers; federal MSFW assistance programs compensate partially for the workers' inability to help themselves.	Narrows and widens current definitions. Nonfarm packing, processing, and fisheries are excluded; livestock included.
2. Eligible farm worker	During the previous 24 months, 50 percent or more of a person's days of work were in qualifying agricultural employment, or 50 percent or more of a person's earnings came from such employment.[35]	Limited resources should be targeted on those who are primarily farm workers. So as to avoid encouraging nonfarm workers to do a little farm work in order to qualify for assistance.	Narrows eligibility for MH and ME; widens eligibility for MHS.
3. Migrant farm worker	Eligible farm workers who moved at least 75 miles and stayed away from home overnight for qualified agricultural employment..	Targets assistance efforts on those whose lives are most disrupted by migration to do farm work. Both the distance moved and overnight stay away from move can be established by intake workers.	Generally narrows the definition of migrant farm worker. Puts greater priority on serving migrants who move the furthest.
4. Duration of eligibility	24 months from last migratory event or qualifying agricultural employment	Best resolves the tension between helping current migrants and farm workers and providing assistance when it is most useful; 24 months is the most frequent duration of eligibility criterion today.	For most programs 24 months requires no adjustment; eligibility for MHS is eased, and is restricted for ME.

NOTES

[1]Varden Fuller, Foreword to Robert D. Emerson, Seasonal Agricultural Labor Markets in the United States, at vii (1984).

[2]Richard Mines and Michael Kearney, The Health of Tulare County Farm Workers, at A1 (mimeo, Apr. 1982).

[3]See Sar A. Levitan, The Great Society's Poor Law: A New Approach to Poverty 251 (1969).

[4]Fuller, supra note 1, at x.

[5]Ronald L. Goldfarb, Migrant Farm Workers: A Caste of Despair 3 (1981).

[6]Fuller, supra note 1, at xi.

[7]As estimated by a Department of Agriculture analysis of supplementary questions attached to the December Current Population Survey (CPS), the number of migrant farm workers rose 21 percent, from 386,000 in 1964 to 466,000 in 1965, before falling to 351,000 in 1966 and then averaging 200,000 during the 1970s. The definition used by USDA to analyze CPS data required persons 14 and older to cross county lines and stay away from home at least one night to be considered migrant farm workers.

[8]For example, during the 1964 debate on what became the Economic Opportunity Act of 1964, the National Sharecroppers Fund Secretary used his understanding that a mechanical lettuce harvester was coming to support his assertion that "machines are replacing men on the farm as they are in the factories." Quoted in Noel H. Klores, Farmworker Programs under the Comprehensive Employment and Training Act – A Legislative History 10 (1981). As of 1993, no lettuce is harvested mechanically in the United States.

[9]The CPS reported an average 883,000 farm workers in 1991, down 23 percent from 1,149,000 in 1983, when women were 25 percent, Hispanics 16, and Blacks 12 percent of all farm workers. These CPS data are drawn from a monthly survey of about 60,000 U.S. households. Statistical Abstract of the United States 394 (1992).

[10]See Philip L. Martin, Harvest of Confusion: Migrant Workers in U.S. Agriculture 74-98 (1988), for an explanation and review of bottom-up estimates of MSFWs.

[11]See id. at 99-109, for an example of a top-down procedure for estimating the distribution of migrant activity across states.

[12]Significant meant one or more counties with 500 or more farm workers or any H-2A temporary foreign agricultural workers (admitted under 8 U.S.C. § 1101(a)(15)(H)(ii)(A) (1988)).

[13]Martin, supra note 10, at 84-88.

[14]In addition, over $50 million has been spent on the MSRTS, which also has a data function for the ME program.

[15]Leslie A. Whitener, Counting Hired Farmworkers: Some Points to Consider (USDA, ERS-AER 524, 1984). Whitener contrasts the characteristics of farm workers in 5 occupational codes from the 1980 Census of Population (475, 476, 477, 479, 484) with characteristics of farm workers in the December 1981 CPS-HFWF sample. According to Whitener, the COP reported that 792,000 farm workers were employed in March 1980, in these occupational codes, while the HFWF estimated 818,000 employed in March 1981. The HFWF indicates that only 1/3 of the hired workers employed sometime during the year are employed in March.

Farm workers employed in the March 1981 HFWF are likely to be year-round workers employed in the South and West and on field crop, livestock, or dairy farms. March workers averaged $7000 in 1981, versus $3500 for workers employed sometime during the year but not in March.

[16]See Martin, supra note 10, at 107-109. Extrapolating from detailed 1984 California Unemployment Insurance (UI) data, there were 600,000 migrant farm workers in the United States if migrant was defined as a worker having at least two farm employers in two counties; 1 million migrant workers if migrant was defined as a worker having a farm job outside the worker's base or highest-earnings county; and 1.2 million migrants if migrant was defined as a worker having a farm job in one county and a farm or nonfarm job in another county.

This top-down procedure relied on national data to distribute farm worker activity across states, and California data to determine the number of farm workers. The UI data are employer-reported worker Social Security numbers (SSNs), and migrant in this top-down study was defined as a worker with an SSN reported by 2 or more employers based in 2 or more counties.

[17]Figure 5.1 is based on quarterly interviews with 7242 farm workers(for some questions) employed in Seasonal Agricultural Services (SAS) between the fall of 1989 and 1991. SAS is most of crop agriculture: it probably includes 80 percent of all farm workers, 70 percent of all farm jobs, and 60 percent of farm wages paid.

[18]Pub. L. No. 99-603, 100 Stat. 3359 (1986). The NAWS was designed to determine whether there were farm labor shortages that would have triggered the admission of "replenishment agricultural workers" under § 303 of IRCA, 8 U.S.C. § 1161 (1988).

[19]Counties were ranked on the basis of the labor expenditures reported by farm employers in the 1987 COA, and QALS employment data were used to determine how many workers to sample in each

region of the United States. See U.S. Dep't of Labor, Findings from the National Agricultural Workers Survey 1990 (Research Report No. 1, 1991).

[20]Victor J. Oliveira, Trends in the Hired Farm Work Force: 1945-87, at 5 (USDA, ERS, Agric. Info. Bull. 561, 1989).

[21]Ed Kissam, A Preliminary Assessment: The 1990 Undercount of America's Farmworkers 11 (mimeo, Mar. 1993).

[22]The NAWS covers SAS agriculture, which by regulation and court decision has been expanded to include most of U.S. crop agriculture. See 7 C.F.R. §§ 1d.9, 1e.2(i) (1992). IRCA states that SAS is to be defined by commodity (perishable) and activity (fieldwork), so that SAW applicants had to be aliens who performed or supervised fieldwork in 1985-86 related to planting, cultural practices, cultivating, growing, and harvesting fruits and vegetables of every kind and "other perishable commodities." The definition of "perishable commodity" was stretched first by USDA and then by courts to include virtually all plants grown for human food (except sugar cane) and many nonedible plants, such as cotton, Christmas trees, cut flowers, and Spanish reeds. Fieldworkers include all of the paid hand- or machine-operator workers involved with these SAS commodities, the supervisors of field workers and equipment operators, mechanics who repair machinery, and pilots who spray crops. These elastic definitions mean that an undocumented Central American refugee paid to work 90 days in a church's vegetable garden could qualify as a SAW applicant. The youngest SAW approved was a 3 year old illegal alien child who helped his parents to bunch onions in 1985-86.

[23]See Findings from the NAWS, supra note 18, at 27. There are many estimates of the percentage of U.S. farm workers who are unauthorized. In hearings and case studies conducted by or for the Commission on Agricultural Workers (CAW) in 1990 and 1991, it was estimated that 10 to 40 percent of the workers were unauthorized. See generally Comm'n on Agricultural Workers, Final Report (Nov. 1992). However, these estimates often refer only to harvest workers, who in the NAWS were only 42 percent of all workers (coincidentally, the percentage of harvest workers is equal to the number of migrant workers in the NAWS). Virtually all harvest workers were aliens, so if 40 percent of all harvest workers were unauthorized, and 20 percent of the remaining alien workers (one-third of the remaining 18 percent of all SAS workers) were unauthorized, then the NAWS is suggesting that less than one-fourth of all SAS workers are unauthorized-- 17 plus 6 percent.

It should be emphasized that many of the unauthorized worker estimates are made by persons who have an incentive to overestimate the percentage. For example, employers who are asked "if IRCA were strictly enforced, would you face a shortage of labor?" have an incentive to overestimate their employment of unauthorized workers, especially if they are also advocating a program which would give them easy access to legal foreign farm workers in the event that unauthorized workers were no longer available.

[24]All crop farms – field crop and FVH – reported $8.2 billion in total labor expenditures in 1987. In the COA, 38 percent of labor expenditures were made by field crop farms, even though the NAWS found that only 7 to 10 percent of the interviewed workers were employed on the field crop farms.

Many of the hired workers on field crop farms whose labor costs are reported to the COA may be artificial in the sense that, if the farm family has a good year, the older children are paid wages to shift net farm income into lower income tax brackets. This shifting of farm income into lower tax brackets may also occur in the CPS, and it helps to explain why farm workers in that survey are distributed in a fashion similar to family farms. Unemployment Insurance or other payroll taxes are typically not paid on these wages.

[25]The 40 percent of the NAWS sample workers who went abroad spent an average 19 weeks there in 1989.

[26]See U.S. Dep't of Labor, California Findings from the National Agricultural Workers Survey (Research Report No. 3, 1993). These data are from the California NAWS interviews conducted between October 1989 and October 1991. The comparisons are not California versus rest of United States comparisons because California data are included as one-fourth of the rest of the U.S. data.

[27]If tree and strawberry production activities each employ 50 or more workers, then in California and many other states, the farm should be reported as two establishments, a berry operation with a SIC 01 code and a tree operation with a SIC 08 code.

[28]See Farmers Reservoir & Irrigation Co. v. McComb, 337 U.S. 755, 762-63 (1949).

[29]29 U.S.C. § 203(f) (1988).

[30]29 U.S.C. § 152(3) (1988). See Bayside Enterprises, Inc. v. NLRB, 429 U.S. 298, 300 & n.6 (1977). The Internal Revenue Code also uses a similar definition of agriculture. 26 U.S.C. § 3121(g) (1988).

[31]Data derived from monthly DOL statistics on employment and earnings.

[32]Nationally, up to 15 percent of total farm wages that are paid to hired workers are paid to relatives of the operator, often in order to shift farm profits into lower income tax categories. This practice seems most common on midwestern grain and livestock farms. Telephone interview with Robert Coltrane, Head, Farm Labor Section, USDA, June 1989.

[33]In 1992, about 3.2 million weeks of UI were paid to persons employed by agricultural reporting units, over one-eighth of all weeks of UI paid in the state. However, UI payments of $338 million were just 9 percent of total UI payments, reflecting lower farm wages.

[34]The cross-county-lines criterion also tends to create more migrants in the eastern states, where counties tend to be smaller.

[35]The average seasonal farm worker is employed about 6 months per year, so this work requirement implies at least 3 months or about 90 days of farm work per year. IRCA required aliens seeking U.S. legal status to have done 90 days of qualifying farm work in the 12 months ending May 1, 1986.

We have not recommended that long season farm workers be ineligible for services. Some MSFWs manage to find work for most of 9 to 10 months, and it seems unfair to deny them access to federal assistance because they found more days of work.

6

The Endless Quest in the 1990s

Producing food and fiber in the United States remains mostly a family affair, but the fastest growing sector of U.S. agriculture is that dependent on hired workers. The value of fruits, vegetables, and horticultural specialties (FVH) is one-sixth of the value of all farm products produced in the United States. U.S. exports of FVH commodities are rising, and today exceed the value of U.S. wheat exports.

In light of the debate over low-wage manufacturing jobs leaving the United States for Latin America and Asia, an expanding FVH agriculture that depends on immigrant workers seems to be out of place. This chapter explores why FVH agriculture seems to defy the economic trends affecting other parts of the U.S. economy, in order to determine whether the demand for immigrant farm workers will persist in the 1990s. Although there are many factors which affect FVH agriculture and thus the demand for migrant workers, three deserve special attention. First, will Americans continue to expand their consumption of the fresh fruits and vegetables that migrants typically harvest? Second, will these fruits and vegetables be produced in the United States or imported? Third, will the fruits and vegetables produced in the United States rely on migrant workers who need federal assistance?

Fruit and Vegetable Consumption

The average American family spends about $28,400 annually, including $4300 on food. About $400 of these annual food expenditures are for the fresh fruits and vegetables that migrants often harvest.[1]

Americans increased their per capita consumption of fruits and vegetables by 10 to 20 percent between 1970 and 1990. The per capita consumption of fresh vegetables during the 1980s, for example, rose 23 percent to 136 pounds per person. This rising per capita consumption was especially noticeable for broccoli and cauliflower; the per capita consumption of fresh broccoli almost tripled from 1.6 to 4.5 pounds during the decade. The consumption of fresh fruit similarly rose sharply during the 1980s, led by increases in per capita consumption of apples, grapes, and strawberries. Despite these increases in per capita consumption, Americans have among the lowest levels of fruit and vegetable consumption among OECD countries (Table 6.1).

The U.S. consumption of most farm commodities increases about 1 percent annually, about the same as the rate of population growth, but the demand for FVH commodities might continue to increase by another 2 or 3 percent annually, or about as much as personal incomes typically go up, because Americans tend to spend about the same percentage more on FVH commodities as their incomes rise. For example, if personal incomes rise 2 percent, then expenditures on broccoli, cauliflower, and similar commodities also rise 2 percent.

Most of the increased demand for fruits and vegetables has been satisfied by U.S. farmers. In 1959, the Census of Agriculture reported that the value of U.S. fruit and nut, vegetable and melon, and horticultural specialty (FVH) production was $2.7 billion, or 9 percent of the total $30 billion of U.S. farm sales. By 1980, FVH sales as estimated by USDA totaled $18 billion, or 13 percent of farm cash receipts, but they jumped by 65 percent during the 1980s to $30 billion in 1990, making these mostly labor-intensive commodities worth about one-sixth of total farm sales.[2] The expansion trends in U.S. fruit and vegetable output are pictured in Figure 6.1.

Even though fruits and vegetables today account for one-sixth of U.S. farm sales and one-quarter of food expenditures, they rarely enter into farm policy debates because the federal government intervenes in these markets in a manner that is not noticeable in federal budget outlays. Instead of guaranteeing minimum prices for fruits and vegetables to prop up grower incomes, the federal government permits fruit and vegetable growers to regulate the manner and, in a few commodities, the volume of what is marketed. The federal government also subsidizes the expansion of the market for fruits and vegetables. U.S. dietary guidelines promote "5-a-day for better health," an attempt to encourage Americans to eat five servings a day of fruits and vegetables.

The federal government purchases fruits and vegetables for child nutrition programs,[3] and it provides $20 billion worth of Food Stamps to poor Americans to buy food, including fruits and vegetables. The federal government has programs to grade and inspect fruits and

Table 6.1. Per Capita Produce Consumption in OECD Countries[a]

Country	Vegetables[b]			Fruit		
	1980	1988	% Change	1980	1988	% Change
	---------------------------Pounds---------------------------					
Canada	328.7	344.8	5	198.2	214.7	8
France	407.9	437.0	7	158.3	177.2	12
Germany	334.0	334.0	0	250.2	256.8	3
Greece	617.1	642.6	4	272.7	242.7	-11
Italy	470.7	469.6	0	268.1	239.6	-11
Japan	314.4	329.8	5	85.8	86.2	0
Spain	571.7	536.8	-6	239.6	222.0	-7
Sweden	255.7	275.9	8	157.0	177.5	13
United Kingdom	405.0	392.9	-3	107.4	124.6	16
United States[c]	302.0	326.3	8	131.0	140.4	7
OECD Average	363.1	380.5	5	185.5	183.5	-1

[a]Includes fresh and processed.

[b]Includes potatoes, sweet potatoes, pulses (beans), fresh vegetables and processing vegetables.

[c]Because it was more complete, the source of U.S. vegetable data is ERS, USDA.
Source: Organization for Economic Cooperation and Development. "OECD Food Consumption Statistics."

vegetables so that they can be more easily marketed, and programs to promote their export : some $60 million—3 percent of the value of fruit and vegetable exports—is provided annually to promote their sale abroad.[4]

All indications are that a growing, affluent, and health-conscious population will continue to increase its fruit and vegetable consumption in the 1990s. Rising consumption during the 1980s translated into an increased demand for migrant workers. Fresh broccoli provides an example. Broccoli is hand-harvested, and its production in California requires an average 52 hours per acre to produce. There was a 50 percent or 40,000 acre increase in U.S. broccoli acreage during the 1980s, so 2.1 million additional hours of labor were needed in the United States to produce broccoli. Even though broccoli is harvested over a long season, enabling workers to average 500 to 1,000 hours annually, the increased production of broccoli, a commodity worth just 1 percent of the total value of FVH commodities, required 2,000 to 4,000 additional seasonal workers just to handle the 1980s increase in production.

Trade and NAFTA

Fruit and vegetable consumption can expand without increasing the number of migrant workers if imports rise. Imports of FVH commodities have been rising – often by dramatic amounts, such as the ten-fold increase in fresh broccoli imports during the 1980s – but from such small bases or starting points that, even after the surge, fresh broccoli imports accounted for only 2 percent of U.S. consumption in 1991. The fresh tomatoes consumed in the United States, a commodity often considered a paradigm of the potential of imports to substitute for U.S. production, are produced primarily in Florida (50 percent), California (25 percent), and Mexico (25 percent). During the 1980s, U.S. consumption, production, and imports of fresh tomatoes each rose by 30 to 40 percent, keeping the import share of the expanding market at about 25 percent. These examples demonstrate the importance of looking at more than just the percentage increase in imports to project the demand for seasonal farm workers in the United States.

Will the North American Free Trade Agreement (NAFTA) shift the production of fruits and vegetables to Mexico? Perhaps over the next decade or two, but not during the 1990s. Mexico's primary competitive advantage is climate; Mexico can produce fresh vegetables during the winter months when most U.S. production areas except Florida are not producing. But even if Mexico completely displaces production in Florida, most fruit and vegetable production will remain in the United States because two-thirds of the production occurs in the summer and fall, when neither Mexico nor Florida is producing significant quantities. But Mexico is unlikely to replace Florida as the source of most winter fruit and vegetables because, in many cases, it is more expensive to produce in Mexico.

For example, one study concluded that in 1991-92, the cost of producing a 25-pound carton of tomatoes was $6.53 in Mexico and $6.40 in Florida. Not only were production costs lower in Florida, Florida tomatoes typically command a $2 per box premium, further reinforcing the Florida advantage. Mexican farm wages are $4 to $6 per day, compared to $4 to $6 per hour in the United States, but Mexican farmers pay proportionately more for worker transportation, housing, and related benefits. Mexican workers in Mexico are less productive (the most productive workers are allegedly in the United States), and the combination of lower productivity, higher non-wage costs, and lower yields combine to make costs of production as high or higher in Mexico.

Mexican worker productivity and yields may increase, but most observers argue that "Mexico's ability to expand horticultural exports depends in large part on attracting foreign investment".[5] Foreign capital is needed especially to finance perennial crop production, such as

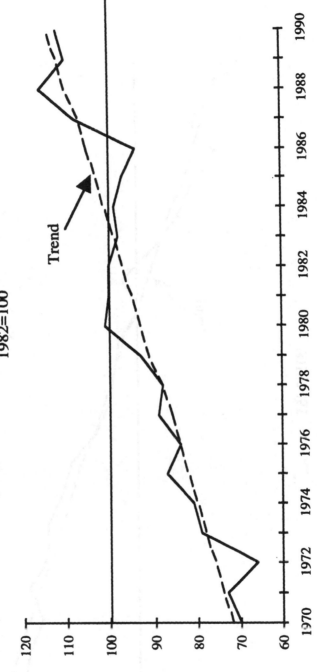

Figure 6.1. U.S. Fruit Output, 1970-1990

1982=100

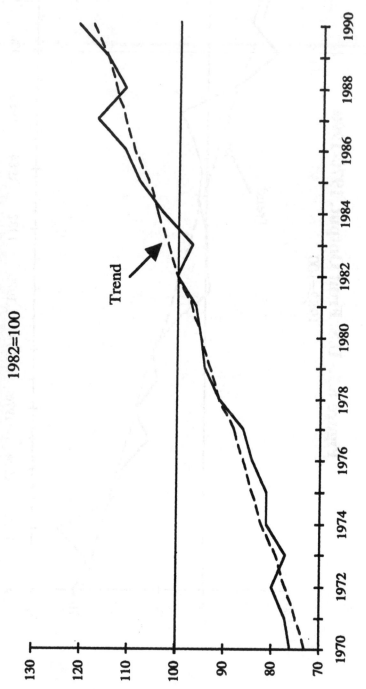

U.S. Vegetable and Melon Output, 1970-1990

1982=100

Trend

Source: USDA

grapes, citrus, and tree fruits, which have a three to seven year lag between the investment and the first crop. Mexico also needs to import the technologies that can improve yields and quality.

U.S. agribusiness today has a relatively small investment in Mexican agriculture: by one estimate, only $30 to $100 million in 1991, or less than California apricot farmers have invested in land to produce apricots. There are several good reasons why U.S. agribusiness has avoided more direct investments in Mexico, including legal restrictions on land ownership. U.S. investors typically produce in Mexico with local partners, and U.S. grower publications are full of stories of U.S. growers who advanced money to their Mexican partners that was never repaid. However, a more important theme of these stories is that Mexican costs of production can be as high as or higher than U.S. costs.[6]

Even if Mexican reforms and NAFTA put U.S. investments in Mexican agriculture on a sound footing, NAFTA is unlikely to spur a shift of production to Mexico. NAFTA eliminates tariffs on fruits and vegetables grown in Mexico. These tariffs are already low -- the average U.S. tariff on the Mexican fruits and vegetables exported to the United States in 1990 was 8 percent. NAFTA eliminates most of these tariffs over 10 years, so that, by 2004, Mexican production costs may fall by 8 percent. But this tariff cost reduction could be dwarfed by the appreciation of the peso expected as NAFTA stimulates foreign investment in Mexico. One study projected a 29 percent increase in the real peso exchange rate due to NAFTA, and such a peso appreciation would raise Mexican production costs far more than tariff reduction lowers the U.S. price of Mexican crops.[7]

NAFTA can lower tariff and non-tariff barriers, but NAFTA will not solve the other problems associated with producing in Mexico for the U.S. market: higher transportation costs, less public research on disease and other factors which reduce yields, and lower worker productivity. These Mexican disadvantages can be overcome with foreign investment and time, but Luis Tellez Kuenkler, Undersecretary in the Mexican Ministry of Agriculture and Water Resources, cautions that Mexican agricultural exports will not skyrocket overnight. He may be correct when he predicted in 1991 that Mexican FVH exports to the United States would rise, at most, by 40 percent within five years after a NAFTA is signed, or from $900 million in 1990 to $1.3 billion by 1998. U.S. FVH production by then is likely to be worth $35 to $40 billion annually, demonstrating that even increased trade with the country best suited to produce for the U.S. market will only marginally slow the expansion of U.S. FVH agriculture, and thus not eliminate the need for migrant workers.

Needy Migrant Workers?

U.S. FVH production could increase in the 1990s to satisfy growing demands for fruits and vegetables in a manner that does not leave the farm workforce in need of federal assistance services. This could happen in several ways. First, the jobs now filled by MSFWs could be eliminated by machines; the equipment operators who replace hand workers typically have fewer needs for farm worker assistance services. Second, the farm labor market could be reformed so that workers earned enough to support themselves and their families. The United Farm Workers (UFW) union was able to bargain for such labor market reforms during the 1970s, enabling some migrant workers to buy homes and settle in the community in which they did most of their work.

Mechanization and collective bargaining depend on the availability of labor – if workers are readily available, there is little incentive to substitute machines for people, and it is hard for unions to bargain for higher wages and benefits. The supply of farm workers, in turn, depends largely on the effectiveness of U.S. immigration controls. Agriculture has long been a major port of entry for unauthorized alien workers, and if such workers continue to arrive in the 1990s, the quest to help needy migrant workers will seem endless.

Mechanization

If fruit and vegetable production continues to expand in the United States, will crops be harvested by Americans operating machinery or by alien workers hand-picking crops? Processing tomatoes is an oft-cited example of what happens to the demand for migrant or immigrant workers after mechanization. In 1960, a peak 45,000 workers (80 percent Braceros) were employed to hand-pick 168,000 acres of tomatoes. Thirty years later, about 5,500 were employed to sort 4 times more tomatoes harvested from 330,000 acres.[8] Growers argued that "the use of Braceros is absolutely essential to the survival of the tomato industry,"[9] but the termination of the Bracero program in 1964 accelerated the mechanization of the harvest in a manner that quadrupled production to 10 million tons between 1960 and 1990.

A uniformly-ripening tomato and a mechanical harvester permitted mechanization to save about 80 percent of the labor needed to hand-pick processing tomatoes. In addition, the tomato harvester changed the work force and the wage system: women paid hourly wages to sort machine-picked tomatoes replaced Bracero men who earned piece rate wages to hand-pick tomatoes.

It is sometimes hard to appreciate the extent to which it was believed in the 1960s that mechanization would soon sweep through fruit and vegetable agriculture, and that farm worker assistance programs were necessary in order to help displaced farm workers to find nonfarm jobs. A 1970 study noted that fruit and vegetable production accounted for 1.1 million jobs in 1968 on 80,000 U.S. farms producing $5 billion worth of fruits and vegetables. Rising wages in the wake of union activities and the end of the Bracero program were expected to accelerate mechanization and displace farm workers, so that, if a crop could not be harvested mechanically, by 1975 it was not expected to be grown in the United States.[10]

The mechanization of the tomato harvest proved to be the exceptional type of labor-displacing change in FVH agriculture, not the rule. There have been important labor savings in FVH agriculture since the 1960s, but they are usually less visible than machines replacing hand harvesters. Changes in production practices for perennial crops have saved labor, such as drip irrigation (which saves irrigator labor), dwarf trees and vines trained for easier hand or mechanical pruning, and precision planting and improved herbicides which save thinning and hoeing labor.

There is also an important counter trend to labor-saving mechanization. Picking and packing grapes, vegetables, and melons in the field increases "farm" employment and reduces "nonfarm" employment because field packing crews both harvest and pack a commodity. The trend toward field packing is uneven, but it has been ascribed to both high wages in unionized packinghouses and portable technologies that make it easier to pick and pack in the fields.[11]

Agricultural engineers note that machines are available to harvest practically every fruit and vegetable grown in the United States, but that machines replace hand-pickers only when it is economically rational to make the switch, or when the cost of machine-harvesting is cheaper than the cost of hand-harvesting. The cost of machine-harvesting falls as technological improvements make machines more efficient, science makes crops more amenable to machine harvesting, and packing and processing facilities become capable of handling machine-harvested produce. Since the technology of hand-harvesting tends to be static, farm wages are the best indicator of the cost of hand-harvesting.

The index of farm wages to the price farmers pay for machinery shows clearly that there was little economic rationale to mechanize during the peak Bracero years (1950-1964). However, after the Bracero program ended in 1964, wages rose faster than machinery prices, prompting the 1960s concern that hand-harvesting jobs would soon be eliminated by machines. Mechanization continued to be a priority for fruit and vegetable growers in the early 1970s, and they supported so many mechanization projects at land-grant universities that the universities were accused of being virtually private research labs for them.[12] However, just as these complaints about growers manipulating university researchers to find machines that would replace troublesome unionized workers reached their peak in the late 1970s, growers lost interest in mechanization.[13] The reason is clear; by the late 1970s, enough unauthorized alien workers were arriving so that using hand-workers was preferred to adopting machines (Figure 6.2).

Farm wages continued to fall relative to the price of machinery throughout the 1980s, reaching a post-war low in 1983-84. Since IRCA was enacted in 1986, the wage-machinery index has risen slightly, but not enough to prompt widespread interest in mechanization. As a result, most of the major fruits and vegetables are still hand-harvested today (Table 6.2).

Farm Unions

The prospects for a revival of unions that can raise farm worker wages are poor. The same factors that slow mechanization – ample supplies of workers that hold down farm wages – also impede effective collective bargaining. The UFW achieved significant wage gains between 1966 and 1980, anchored by 40 percent one-year wage increases in the first and last years. But these union collective bargaining victories had different origins, and ushered in eras of sharply different prospects for further union gains.

The UFW won a 40 percent wage gain for grape harvest workers in 1966 by mounting a boycott of the liquor produced by a conglomerate that also grew table grapes. Table grapes were a very small part of the target company's business, and grapes were being produced as much to profit from rising land prices as to sell grapes. The union could persuade the company to raise wages from $1.25 to $1.75 hourly – at a time when the federal minimum wage was $1.25 hourly – because most large growers seemed resigned to paying higher wages in order to obtain farm workers now that Bracero workers were no longer available.

The UFW's fortunes rose and fell several times after it burst on the scene as a private self-help organization for farm workers. Its high-water mark came in 1975, when the California legislature enacted the Agricultural Labor Relations Act (ALRA), a state law that filled the void created by the NLRA's exclusion of farm workers. The ALRA has been described as the most pro-worker and pro-union labor relations law in the United States.[14] Its features include a makewhole provision under which employers who fail to bargain with a certified union bargaining

Table 6.2. Percentage of Major Fruits and Vegetables That Are Hand-Harvested, 1990

	Fruits[a]		Vegetables[b]	
Acreage hand harvested (%)	Commodity	Cash receipts (1990$mil)	Commodity	Cash receipts (1990$mil)
76-100	Apple	1,445	Cantaloupe	182
	Grape*	1,662	Lettuce	847
	Peach*	365	Asparagus	148
	Pear*	262	Cauliflower	190
	Strawberry	594	Peppers	219
	Apricot	41	Broccoli*	268
	Sweet cherry*	119	Celery	214
	Grapefruit*	384	Cabbage	121
	Lemon*	278	Cucumber*	215
	Plums and Prunes*	267	Mushrooms	502
	Orange*	1,455	Watermelon	126
			Honeydews	82
51-75			Sweet potatoes	108
26-50			Onions**	551
			Tomatoes*	1,622
0-25	Tart cherry*	37	Carrots	273
	Cranberry	153	Sweet corn*	467
			Dry Beans	675
			Potatoes	2,678
			Peas*	132
			Lima beans*	--
			Snap beans*	204
	Subtotal	7,062	Subtotal	9,416

[a]These major fruits accounted for 92 percent of the $7.7 billion value of all fruit sales in 1990.
[b] U.S. cash receipts for vegetables and melons were $11.5 billion in 1990; these major vegetables accounted for 80 percent of these receipts.
* More than 50 percent of crop is processed.
**Includes hand-harvested green onions

Source: U.S. Department of Agriculture.

representative in good faith can be ordered to make their workers whole for any loss of wages and benefits they suffered. In 1975, the average entry-level wage on unionized farms in California was $3.11 per hour,[15] so that a worker employed full-time (2,000 hours) could earn more than the poverty-level income for a family of four ($5,500).

In 1979, when this first round of contracts signed under the ALRA was expiring, the UFW demanded another 40 percent wage increase. Growers resisted. The union called strikes that reduced the supply of lettuce by about one-third, but the strike boomeranged and raised total grower revenues because prices tripled. Some growers were able to harvest their crops with crews of unauthorized workers assembled by farm labor contractors (FLCs). With immigration reducing the effectiveness of the strike, the UFW mounted boycotts against the

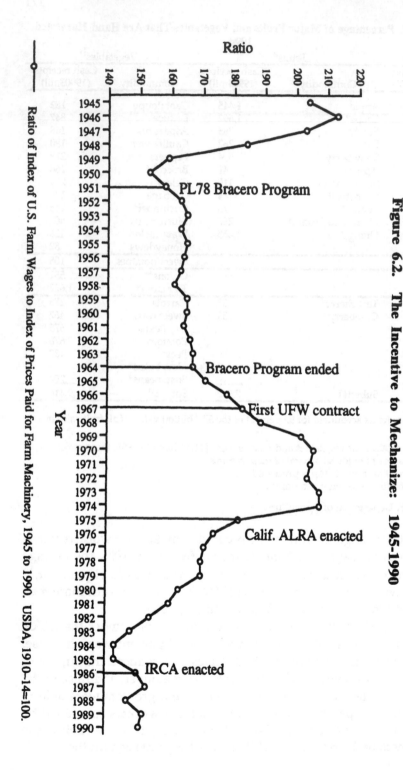

Figure 6.2. The Incentive to Mechanize: 1945-1990

Ratio of Index of U.S. Farm Wages to Index of Prices Paid for Farm Machinery, 1945 to 1990. USDA, 1910-14=100.

nonfarm products of several of the corporate growers that also raised vegetables.

The UFW eventually won the 40 percent one-year wage increase it demanded, raising entry-level wages from $3.75 to $5.25 hourly (the federal minimum wage was $3.10 hourly in 1980), but this wage gain proved to be a Pyhrric victory. Many of the corporate growers signed UFW contracts in order to go out of business, and others switched from being both growers and packers to being only nonfarm packers. Thereafter the smaller farmers who grew lettuce or broccoli assumed the responsibility for getting it harvested. Union membership plummeted; by some reports, from 60,000 to 90,000 in the early 1980s to 6,000 to 9,000 a decade later.[16]

The UFW successfully showed that, under conditions of labor scarcity and favorable legal circumstances, farm workers could help themselves. In 1978, when a poverty-line income was less than $6,700, many UFW members earned considerably more. In addition, the UFW operated an employer-paid health insurance program, and required employers to make pension contributions on the workers' behalf.

The UFW unraveled in the 1980s for a number of reasons, including internal problems (Cesar Chavez forced out most of the non-Hispanic lawyers and organizers), more sophisticated and determined employer resistance, and changes in the administration of the ALRA that followed the election of a Republican governor in California in 1982.

The most important factor in the demise of UFW's self-help effort was the upsurge in unauthorized immigration. Studies of labor markets that were transformed from union to non-union status revealed a common pattern. The UFW would call a strike in support of its demand for a wage increase above the then common $5 per hour level; growers would turn to FLCs to obtain replacement crews that included large numbers of unauthorized workers; and the strike would end with the UFW filing a barrage of bad-faith bargaining charges against the employer. Many UFW members would work at the lower wages introduced by FLCs to survive, in the hope that they would be fully compensated eventually by the state agency that administered the ALRA. They were often disappointed, especially after a California court interpreted the law to mean that, even if the employer did bargain in bad faith with the union, if the employer could show that even good faith bargaining would not have resulted in higher wages, makewhole wages and benefits need not be paid.[17] Many growers pointed to the wages and benefits paid by FLCs and were found to owe nothing.

There are other unions active in California, but none has so far repeated the UFW's success of the late 1970s in raising wages enough so that a farm worker family can earn more than a poverty-level income ($13,400 in 1990). The recent immigration of Indian populations such as

Mixtecs from southern Mexico has spawned self-help organizations, including several union election victories, but these organizations have not yet won wage and benefit increases. In fact, the typical issue for unions with farm worker members today is to resist wage and benefit rollbacks.

Self-help union efforts in other states have also been frustrated by an oversupply of workers. The most successful self-help effort recently has been in Ohio, where the Farm Labor Organizing Committee (FLOC) has a three-way contract between food processors, growers, and workers. While this boycott-inspired agreement raised wages to workers, as well as prices to growers so that they could pay these higher wages, it has not lifted most of the migrant workers covered above the poverty income threshold. Instead, many of the cucumber harvesters covered by the FLOC agreement utilize farm worker service programs.

Continuing Immigration

So long as unauthorized immigration continues, there is likely to be little mechanization that eliminates the need for migrant workers and few successful self-help union efforts that make farm worker service programs unnecessary. What are the prospects for continuing unauthorized immigration?

The United States enacted the Immigration Reform and Control Act (IRCA) of 1986 to reduce illegal immigration by making it more difficult for unauthorized workers to find U.S. jobs, and to offer a legal immigration status to certain unauthorized aliens. IRCA included three major agricultural provisions: deferred sanctions enforcement[18] and search warrants, the SAW legalization program, and the H-2A and the RAW foreign worker programs.[19] Although IRCA was meant to control illegal entries, it probably facilitated such entries, especially for farm workers.

Until IRCA, INS enforcement in agriculture often involved the Border Patrol driving into fields and apprehending aliens who tried to run away. Farmers pointed out that the INS was usually required to obtain search warrants before inspecting factories for illegal aliens, and argued that the INS should similarly be obliged to show evidence that illegal aliens were employed on a farm before raiding it. IRCA extended the requirement that the INS have a search warrant before raiding a workplace for illegal aliens from nonfarm to agricultural workplaces.[20]

IRCA created two legalization programs: a general program which granted legal status to illegal aliens if they had continuously resided in the U.S. since 1982,[21] and the SAW (seasonal agricultural worker) program, which granted legal status to illegal aliens who did at least 90 days of farm work in 1985-86.[22] Because farmers and farm worker advocates testified that many illegal alien workers were paid in cash, Congress made it easier for SAW applicants to prove that they satisfied this work requirement than it was for nonfarm aliens to prove that they met the residence requirement. A SAW applicant, for example, could have entered the United States illegally in early 1986, left after doing 90 days of farm work, and then applied for SAW status from abroad. An applicant was permitted to apply for the SAW program with only an affidavit from an employer asserting that the worker named in the letter had done, e.g., 90 days of work in virtually any crop. The burden of proof then shifted to the Immigration and Naturalization Service (INS) to overcome the alien's showing of claimed employment.[23]

Almost 1.3 million aliens applied for legal status under the SAW program, far more than anyone expected. Most observers had put the number of persons who did 90 or more days work on crop farms at 1 to 1.5 million, and estimates of the percentage of them who were unauthorized ranged from 10 to 40 percent.[24] It was on the basis of such guesses that the SAW program was divided into two parts: Group I SAWs had to do at least 90 days of qualifying work in each of the years ending May 1 of 1984, 1985, and 1986, but a maximum 350,000 such workers could be in this group. They could convert from temporary to permanent resident alien status a year earlier than Group II SAWs, those who could show 90 days of qualifying work only in the year ending May 1, 1986. This 350,000 figure was a consensus estimate of the number of unauthorized aliens eligible for the SAW program.

The SAW program turned out to be a case of good intentions gone awry in the effort to restrict the entry of alien farm workers. Fears of labor shortages during the spring of 1987, symbolized by an early bumper crop of strawberries and grower complaints of labor shortages, prompted California growers to get Congress to pressure the INS to accept skeletal applications for SAW status at ports of entry. According to growers, there were thousands of workers eligible for SAW status in Mexico, but the only proof of their qualifying employment was with a U.S. employer whom they would recognize, but for whom they might not have a name or address.

As word spread that the INS was permitting farm workers to enter the United States legally and look for the records needed to apply for SAW status, as well as granting 90-day work permits to such entrants, many rural Mexicans set off for the United States.

SAW applicants turned out to be mostly young Mexican men (Table 6.3). (There were only about 6 million adult men in rural Mexico in 1987 and 1988; the equivalent of one-sixth of them applied for SAW status.) Their median age was 24, and half were between 20 and 29. Since SAWs had to be employed in 1985-86 to qualify, there were few SAWs under

15, compared to 7 percent of the general legalization applicants. Over 80 percent of all SAW applicants were male, and 42 percent were married. In a few limited surveys, SAWs who had an average 5 years of education earned between $30 and $35 daily for 100 days of farm work in 1985-86.

The INS was aware of the abuses to which the SAW program was subject, such as coaches who rented farm worker clothing to nonfarm workers near U.S. ports of entry. But Congress had designed the program, at the growers' urging, to favor admitting rather than excluding applicants. Most SAW applicants submitted only a short letter from a U.S. employer which asserted that the alien had done, e.g., 91 or 92 days of work picking tomatoes in 1985. The burden of proof then shifted to the INS. Although the INS did expose several FLCs who sold thousands of false work histories, the INS was not equipped to reject applications that, e.g., claimed 90 days of harvesting work in a crop whose harvest is at most 60 days.[25] In one of the many post-mortems of the easy SAW legalization program, Robert Suro of the New York Times described it as "one of the most extensive immigration frauds ever perpetrated against the U.S. government."[26]

IRCA strengthened rather than weakened the links between rural Mexico and rural America. Instead of discouraging Mexicans from seeking U.S. farm jobs, IRCA made it easier to find employment in the United States. Of course, the other part of IRCA, employer sanctions, was supposed to have dried up the employment opportunities for future unauthorized aliens (those who had not gained legal status under the SAW or general legalization programs). But it has become clear that the sanctions program can be rendered ineffective by bogus documents that are not hard to acquire. Congress attempted to step up the battle against such schemes in 1990, by introducing civil money penalties for knowingly using fraudulent documents to obtain a U.S. job,[27] but the provision is, at best, being slowly implemented. IRCA therefore converted enforcement in agriculture from a sometimes disruptive pursuit of aliens through the fields to a paper chase that does not seem to disrupt many farm operations.

IRCA is likely to be remembered as a stimulus to unauthorized immigration. Instead of ushering in an era in which legal workers could mount self-help efforts that might raise farm wages and perhaps spur mechanization, IRCA unleashed a new wave of unauthorized immigration that promises new challenges to farm worker assistance programs.

There are few indications that immigration laws will be enforced in the 1990s in a manner that breaks these cross-border labor market links. This is most unfortunate, because controlling illegal migration probably offers the single most effective strategy for introducing steady long-term

Table 6.3. General and SAW Legalization Applicants

Characteristic	General or Pre-1982 (a)	SAW (b)
Median Age at Entry	23	24
Age 15 to 44 (%)	80	93
Male (%)	57	82
Married (%)	41	42
From Mexico (%)	70	82
Applied in California (%)	54	52
Total Applicants	1,759,705	1,272,143

Source: INS Statistical Yearbook, 1991, pp. 70-74
(a)Persons filing I-687 legalization applications
(b)Persons filing I-700 legalization applications. An additional 80,000 farm workers received legal status under the general or pre-1982 legalization program.

improvement in the lives and working conditions of farm workers. Assistance programs by themselves at best can mitigate the hardships of life as experienced by MSFWs and their families, and can help some individuals to escape into nonfarm employment. But long-term alleviation of these conditions requires better wages, better working conditions, and more effective enforcement of protective legislation (such as wages and hours requirements or housing code restrictions). All of these aims are undercut as long as a new and more docile cheap labor force remains readily available via undocumented migration. And as we have seen, the same undocumented migration retards mechanization and undermines unionization efforts.

Farm worker advocates tend to align themselves with opponents of IRCA's employer sanctions, and they would certainly feel uncomfortable climbing aboard the bandwagon with the typical supporters of tougher immigration law enforcement. We share some of this apprehension. But as we have tried to show, the other means apparently available for genuinely improving farm worker welfare are badly hampered by the continuing availability of new undocumented workers. True concern for this historically underappreciated and exploited group of workers has to be manifested in an honest recognition of these counterintuitive insights. If assistance workers want to stop playing Sisyphus, they need to support better controls over illegal migration.

Whither Farm Worker Assistance?

The federal government launched the Big 4 farm worker assistance programs in the 1960s, when "every one knew" how to identify the farm workers who needed help, and government officials, service providers, farmers, and academics agreed that the goal of federal service programs was to help soon-to-be-displaced farm workers to find nonfarm jobs.

Farm labor events did not unfold as anticipated. In some ways, farm worker assistance programs were enormously successful. The U.S. citizen white and black migrant workers of a generation ago have largely disappeared, and many of these workers and their children made the transition to nonfarm jobs with the help of farm worker assistance programs. But the vacuum they left in the farm labor market was filled by even needier immigrants who arrived in such large numbers, and in such a vulnerable position, that wages, mechanization, and self-help efforts stagnated.

Farm worker assistance programs have made uneven progress to reach their new clientele. We have urged them to take steps to improve coordination between programs and to lay the groundwork for better data collection. But we found remarkably few service providers asking the big question of whether their Sisyphean task makes sense. Should the U.S. government continue to spend the equivalent of 10 percent of the annual earnings of migrant farm workers to improve the lives of some of them? Shouldn't it focus instead on harnessing labor market forces to help improve the lives of all of them?

We were impressed by the scores of people we met in the course of this study who have dedicated their working lives to the welfare of farm workers and their families, but we think they are currently engaged in an endless quest. Success usually means one family's escape from the farm labor market and the entry of a new and often needier immigrant worker. Even if federal funding for farm worker assistance programs were doubled, many farm worker families would remain poor and in need of health and education services. For this reason, we urge serious consideration of efforts to reform the farm labor market so that workers employed there do not need special federal assistance programs.

The key determinant of farm worker welfare is the supply of workers – a tight labor market is a farm worker's best friend. The farm labor market today is awash with immigrant workers because of an immigration reform gone awry, holding down farm wages and dooming self-help efforts. The federal government in the mid-1960s launched the closest thing we have had to a golden era for farm workers by ending the Bracero program; it should launch another in the 1990s by enforcing U.S. immigration laws.

NOTES

[1]The data in this section are drawn from USDA reports on fruit and vegetable production and consumption. USDA reported that fruits and vegetables were 23 percent of the total $441 billion that consumers spent on retail food items in 1990, but USDA includes processed fruits and vegetables, as well as baby food, dressings and condiments, in this broad fruit and vegetable category.

[2]These data (COA production data and ERS sales data) are not strictly comparable. The 1987 Census of Agriculture reported that FVH sales were $17.6 billion, or 13 percent of farm cash receipts.

[3]Section 32 of the Agricultural Adjustment Act of 1935 authorizes an appropriation equal to 30 percent of the U.S. customs receipts from the prior calendar year to encourage the domestic consumption of farm products, to promote their export, and to raise farm incomes. These funds ($4.7 billion in FY 91) are used to pay for most of the cost of USDA's child nutrition programs. Funds remaining after these program costs are covered are used to purchase "perishable" farm products; in FY 90, $128 million or one-third of the $379 spent on direct purchases was used to buy mostly processed fruits and vegetables.

[4]Between 1986 and 1990, $271 million was provided to promote fruits and vegetables through the Targeted Export Assistance Program. Commodities receiving assistance included California and Florida citrus, which received $83 million or 30 percent of all TEA funds. California raisins received $49 million, or almost 20 percent of TEA funds. Both citrus and raisins are picked mostly by recent Mexican immigrants.

[5]Cook, Roberta et al. Fruit and Vegetable Issues, Vol. 4 of NAFTA: Effects on Agriculture (1991).

[6]See the articles in The Packer, April 25, 1992, at 1A et seq.

[7]Gary Hufbauer & Jeffrey Schott, North American Free Trade: Issues and Recommendations 57 (1992).

[8]It is very difficult to find consistent data on farmworker employment by commodity. A common approach is to ask farm advisors to estimate the hours of regular and seasonal labor required per acre to produce a commodity, but there are often unexplained differences between reports for the same commodity that is produced similarly in 2 counties. One farm advisor reported that the 40,000 acres of processing tomatoes in Yolo county required 6 regular and 38 temporary hours of labor per acre in 1989, while the 63,000 acres in Fresno county required 22 regular and 31 seasonal hours per acre. These numbers were combined in Employment Development Dep't of California, Seasonal Labor in California Agriculture: Labor Inputs for California Crops, Agricultural Studies Report 90-6 (John Mamer & Alex

Wilke, 1990) to generate a statewide average 16 regular and 34 seasonal hours per acre.

The major reason for this difference in hours of labor needed to produce processing tomatoes is that in Yolo county, irrigation hours were reported to be 0, while in Fresno, they were reported to be 7 hours per acre. The Fresno report also included 5 hours of regular supervisory labor per acre, while the Yolo report had none.

Both county reports estimated that harvesting required 11 to 13 hours per acre, or that harvesting required fewer hours per acre than thinning and weeding (14 hours). At 12 hours per acre, sorting the tomatoes from 330,000 acres required 4 million hours of labor. Sorters sometimes work during peak season 12 hours per day and 6 day weeks; if they average 72 hours for 10 weeks, they average 720 hours per season. These calculations suggest that a total 5500 sorters each working 720 hours would be required to harvest California's processing tomatoes.

[9.]Cal. Senate Fact Finding Comm. on Labor and Welfare, California's Farm Labor Problems: Part I, at 105 (1961).

[10]Fruit and Vegetable Harvest Mechanization (B. F. Cargill & G. E. Rossmiller eds., 1970).

[11]The alternative to field-packing is to have farm workers pick a crop such as broccoli into bulk bins, haul it to a packingshed, and have nonfarm shed workers sort and bunch it. This distinction is important under the NLRA, FLSA, and IRCA. If the packingshed primarily packs broccoli grown by one farmer, the shed workers are "farm workers," and thus excluded from overtime laws under the FLSA. Under IRCA, any undocumented aliens involved may have been eligible for the SAW program.

If the packingshed is "commercial," i.e., packing broccoli for several farmers, then the shed workers are nonfarm workers who must be paid premium overtime wages. They would not have been for the SAW program.

[12]See James Hightower, Hard Tomatoes, Hard Times (1978). But see Philip Martin & Alan Olmsted, The Agricultural Mechanization Controversy, Science, February 8, 1985, at 601-606.

[13]G.K. Brown, Fruit and Vegetable Mechanization, *in* Migrant Labor in Agriculture: An International Comparison at 200 (Philip Martin ed., 1984).

[14]Karen Koziara, Agriculture, *in* Collective Bargaining: Contemporary American Experience 263-314 (G. Somers ed., 1980).

[15]The federal minimum wage was $2.10 hourly in 1975.

[16]Philip Martin, Suzanne Vaupel, and Daniel Egan, Unfulfilled Promise: Collective Bargaining in California Agriculture (1988).

[17]Philip Martin and Daniel Egan, The Makewhole Remedy in California Agriculture, 43 Industrial & Labor Relations Rev. at 120-130 (Oct. 1989).

[18]8 U.S.C. § 1324a(i)(3) (1988) delayed enforcement of sanctions against most agricultural employers for about 18 months, in comparison with other employers.

[19] The H-2A program for temporary admission of agricultural workers, which had existed for many years, was refined and more closely regulated under IRCA. 8 U.S.C. § 1188 (1988). The replenishment agricultural worker (RAW) program was enacted at growers' insistence to provide for new permanent admissions in FY 1990-93 if labor shortages developed in the wake of IRCA enforcement. 8 U.S.C. § 1161 (1988). There were no such shortages, so RAW workers were never admitted.

[20]8 U.S.C. § 1357(e) (1988).

[21]8 U.S.C. § 1255a (1988).

[22]8 U.S.C. § 1160 (1988).

[23]See id., § 1160(b)(3)(B)(iii).

[24] For a discussion of these estimates, see Philip L.Martin, The Outlook for Agricultural Labor in the 1990's, 23 U.C. Davis Law Review (Spring 1990).

[25]See Philip Martin, Edward Taylor, and Philip Hardiman, California Farm Workers and the SAW Legalization Program, California Agriculture, November-December 1988, at 4-6.

[26]Reprinted in the Sacramento Bee, November 12, 1989, at A1.

[27] 8 U.S.C. §1324c (Supp. 1991).

Appendix A. Recommendation 92-4 of the Administrative Conference of the United States 1 C.F.R. §305.92-4 (1993)

Adopted June 19, 1992

Coordination of Migrant and Seasonal Farmworker Service Programs

Since the 1960s, the federal government has established numerous service programs to help meet the needs of migrant farmworkers. From the early days, migrants have been considered a uniquely federal responsibility, primarily because of their interstate movement, which makes it hard for the workers and their families to qualify for local assistance and disrupts other services like schooling for the children. As these programs have evolved, many have come to serve nonmigrant seasonal farmworkers as well.

The programs to meet health, education, housing, job training, and other needs of migrant and seasonal farmworkers (MSFWs) have developed separately. There are approximately 10 MSFW-specific service programs, and farmworkers also draw upon the assistance of numerous other general programs such as food stamps or Medicaid. The four largest federal programs are Migrant Education, administered by the Department of Education; Migrant Health and Migrant Head Start, both administered by the Department of Health and Human Services; and the Department of Labor's special job training programs for MSFWs under Section 402 of the Job Training Partnership Act.

Each program has its own definition of migrant and/or seasonal farmworker, as well as other eligibility standards. The result is a potential for overlap of some services and gaps in others, and there is no overarching provision for effective coordination among the programs. Various efforts have been undertaken at the national level to improve coordination, with mixed success to date. These include an Interagency Committee on Migrants, a staff-level group that meets quarterly, largely for information-sharing purposes; an Interagency Coordinating Council, established informally as a forum for policy-level decisionmakers involved in the various programs, but now inactive; and a Migrant Inter-Association Coordinating Committee, involving nonprofit grantees and other organizations representing direct service providers.

In addition, MSFWs often qualify for other services provided by state and local governments or funded through private initiative, each governed by its own particular definitions or eligibility standards. These services are especially important in areas where some or all of the major federal programs are not present. Effective local service providers therefore have to be adroit in locating those available services, from whatever source, that can best meet the needs of their clientele. Because of the great variety in locally available services of this kind, much of the task of coordination among MSFW service programs necessarily takes place at the local and state level. Many states are finding ways to encourage this process by the creation of a governor's committee or task force, involving service providers, growers, representative government officials, farmworkers, and others.

The federal government should also take steps to improve coordination of services. For example, the intake procedures for each service program (now typically undertaken separately by each of the agencies, despite considerable duplication) should be streamlined. To effectuate such efforts, and to provide better interagency consultations before program changes are introduced, the President should establish by executive order a policy-level Interagency Coordinating Council on MSFW programs. This Council is not intended to replace, and indeed should promote, existing coordination at the program staff, state, and service delivery level.

To facilitate interagency coordination, whether or not such a Council is created, a reliable system for gathering data on the nation's population of MSFWs is needed. Although each agency has its own mechanism for generating program statistics and estimates of the target population, these vary widely in method and scope, and each suffers from specific inadequacies. They produce widely varying pictures of the nation's population of MSFWs, to the continuing frustration of legislators, service providers, researchers, and others. Agricultural labor data have always been left out of the Department of Labor's regular employment data system, and no other adequate permanent data source now fills the gap. The recommendation provides some guidance on the goals of such an information-gathering effort.

Recommendation

I. *Coordination at the National Level*

An Interagency Coordinating Council on migrant and seasonal farmworker (MSFW) programs should be established to strengthen national coordination of MSFW service programs. The Council would be charged, inter alia, with identifying specific coordination tasks to be accomplished, in most cases under the primary responsibility of a designated lead agency.

A. To ensure an enduring structure and a clear mandate, the President should issue an executive order creating the Council, specifying the policy-level officials from appropriate agencies who would be permanent members and designating a chair. The order should also designate an agency that would initially have primary responsibility for staffing the Council's meetings and other functions. The Council should be specifically charged to coordinate the review MSFW service programs, giving particular attention to gaps in services and unjustified overlap. It should encourage public participation through public meetings, creation of an advisory committee, or other means.

B. The executive order should provide that the Council, in cooperation with the Office of Management and Budget, review proposals for significant changes in any agency's MSFW service program (including proposed legislation, regulations, and grantee performance standards). OMB should consolidate or coordinate its own oversight of all federal MSFW service programs.

C. The executive order should assign to the Council the initial responsibility to develop, through delegations to the appropriate agencies, a reliable and comprehensive MSFW population census system, independent of any of the specific programs, along the lines described in part II. Other specific coordination tasks that the Council might wish to take up include development of consolidated or streamlined intake processing for MSFW programs, provision of better linkages among existing MSFW information clearinghouses, and encouragement of cooperation among direct service providers.

D. The Council should identify and assign priorities to the coordination tasks to be accomplished, with a strategy and timetable for their achievement. In most instances, it should assign lead responsibility for each specific coordination task to a designated agency. That agency's coordination efforts with other agencies may include suggesting regulations or other implementation measures.

E. The Council should study the differing eligiblity standards of MSFW programs and identify, if appropriate, where consistency could be achieved without substantial impact on the beneficiaries of those programs.

F. The Council should also study and make recommendations on the strengthening of state and local coordination of MSFW programs.

II. *Information Gathering on Migrant and Seasonal Farmworkers*

A. To improve coordination of and service delivery in MSFW programs, the executive order should:

(1) Authorize the Council to develop an integrated, cost-effective system for gathering data on the number, characteristics, and distribution of MSFWs and their dependents;

(2) Authorize the Council to designate an appropriate agency to have responsibility for collecting the data, with the cooperation of federal agencies with MSFW service programs;

(3) Direct appropriate federal agencies with expertise in gathering these kinds of data, such as the Bureau of the Census, the Bureau of Labor Statistics, the National Center for Education Statistics, or the National Agricultural Statistics Service, to cooperate with the Council's effort; and

(4) Provide opportunities for submission of data and information from the public.

B. This data system should ensure that the information gathered on MSFWs and their dependents sufficiently describes workers employed in a broad spectrum of U.S. agriculture and related industry. This means that the data should include and distinguish among workers employed, for example, in crop and livestock production, the packing and processing of farm products, and fisheries. Data should be collected on workers and their dependents, including such factors as recency and frequency of migration, farm and nonfarm earnings and periods of employment, and health, education and housing characteristics. These comprehensive data should be collected in a form designed to be useful to service programs with differing definitions of eligible workers and their dependents.

C. This data system should be designed to help the Council identify general trends—including changes in the total number of MSFWs and their dependents and employment patterns—and opportunities for coordination among MSFW programs. To help achieve this goal, the Council should consider whether there are areas in which a consensus on a set of common characteristics of MSFWs should be developed for statistical purposes.

Appendix B. Migrant Services Directory
(October 1992)

USDA, FARMERS HOME ADMINISTRATION

COMMUNITY FACILITIES LOAN PROGRAM

Purpose: The Community Facilities Loan Program provides loans to public bodies, nonprofit organizations and Federally recognized Indian tribes in rural areas of 20,000 or less in population for essential community facilities projects involving health care, public safety, public service, education and cultural activities.

Contact:
James C. Alsop
Branch Chief, Program Development
Community Facilities Division
14th & Independence Ave., S.W., Room 6304-S
Washington, DC 20250
Telephone: (202) 720-1497
FAX: (202) 690-3808

FARM LABOR HOUSING LOAN AND GRANT PROGRAM

Purpose: The Farm Labor Housing Loan and Grant Program provides funds for the construction of housing and other ancillary facilities located at a housing site.

Contact:
John Pentecost, Branch Chief
Multi-Family Housing Division
Farmers Home Administration, USDA
14th and Independence Ave., S.W., Room 5343-S
Washington, DC 20250
Telephone: (202) 720-1606
FAX: (202) 720-0302

USDA, FOOD AND NUTRITION SERVICE

CHILD AND ADULT FOOD CARE PROGRAM

Purpose: The Child and Adult Care Food Program provides cash reimbursement and donated foods to help provide nutritious meals and snacks to children enrolled in nonresidential child care centers or day care homes.

Contact:
Robert M. Eadie, Chief
Policy and Program Development Branch
Child Nutrition Programs
Food and Nutrition Service, USDA
3101 Park Center Drive, Room 1007
Alexandria, VA 22302
Telephone: (703) 305-2619
FAX: (703) 305-2879

NATIONAL SCHOOL LUNCH, SCHOOL BREAKFAST AND SPECIAL MILK PROGRAMS

Purpose: The National School Lunch Program provides subsidized lunch service for eligible school children. The School Breakfast program provides subsidized breakfast service for eligible school children. The Special Milk Program provides subsidized or free 1/2 pints of milk for children in participating private and public schools.

THE SUMMER FOOD SERVICE PROGRAM

Purpose: The Summer Food Service Program provides cash reimbursement and donated foods to help provide nutritious meals and snacks to children from needy areas during the summer.

SPECIAL SUPPLEMENTAL FOOD PROGRAM FOR WOMEN, INFANTS AND CHILDREN (WIC)

Purpose: The WIC Program provides specific nutritious supplemental foods and nutrition education to low income pregnant, postpartum and breast feeding women. It serves infants and children up to age five who are determined to be at nutritional risk. Special program provisions have been implemented in response to the unique needs of the migrant population.

WIC legislation provides for a special migrant set aside of not less than nine-tenths of one percent (.9%) of appropriated funds for services to eligible migrant populations.

Contact:
Ronald J. Vogel, Director
Supplemental Food Programs
Food and Nutrition Service,
USDA
3101 Park Center Drive, Room 540
Alexandria, VA 22302
Telephone: (703) 305-2746
FAX: (703) 305-2420

DEPARTMENT OF EDUCATION

BILINGUAL EDUCATION PROGRAMS

Purpose: The Office of Bilingual Education and Minority Languages Affairs sponsors programs at various levels to assist students with limited English proficiency, including: a State Education Agency Program; the Transitional Bilingual Education (TBE) Program; the Special Alternative Instructional Program; the Special Populations Program; the Educational Personnel Training Program; the Training, Development and Improvement Program; and the National Clearinghouse for Bilingual Education.

Contact:
Rebecca Richey,
Program Specialist
OBEMLA
Department of Education
Switzer Bldg., Room 5086
330 C Street, S.W.
Washington, DC 20202-6641
Telephone: (202) 205-9717
FAX: (202) 205-8737

CHAPTER 1 AND RELATED PROGRAMS

Purpose: Chapter 1 provides funds to establish and improve programs for all disadvantaged children and at-risk youth, including the children of migratory agricultural workers and fishers.

Contact:
Mary Jean LeTendre, Director
Compensatory Education
Department of Education
400 Maryland Ave., S.W.,
Room 2043
Washington, DC 20202
Telephone: (202) 401-1682
FAX: (202) 401-2527

CHAPTER 1 TECHNICAL ASSISTANCE CENTERS (TACs) AND RURAL TECHNICAL ASSISTANCE CENTERS (RTACs)

Purpose: These projects provide technical assistance to States and localities on a variety of issues concerning Chapter 1 Migrant Education Programs, regulations and evaluation. The TACs and RTACs work to improve Chapter 1 projects, evaluate program effectiveness and assist States in implementing Chapter 1 program changes.

Contact:
Daphne Hardcastle,
Coordinator
Chpt. 1 TACs and RTACs
Planning and Evaluation
Department of Education
400 Maryland Ave., S.W.,
Room 3127
Washington, DC 20202
Telephone: (202) 401-1958
FAX: (202) 401-3036

HANDICAPPED MIGRATORY AGRICULTURAL AND SEASONAL FARMWORKER VOCATIONAL AND REHABILITATION SERVICES PROGRAM

Purpose: Projects funded by this program provide vocational rehabilitation services to handicapped migratory and seasonal farm workers and services to family members who are with them.

Contact:
Edward Hofler, Program Officer
Rehabilitation Services
Administration
Department of Education
Switzer Building, Room 3318
330 C Street, S.W.
Washington, DC 20202-2740
Telephone: (202) 205-9432
FAX: (202) 205-9772

MIGRANT EDUCATION PROGRAMS

Contact:
Francis V. Corrigan, Director
Office of Migrant Education
Department of Education
400 Maryland Ave., S.W.,
Room 2149
Washington, DC 20202-6135
Telephone: (202) 401-0740
FAX: (202) 401-2528

The programs described below fall under the purview of the Migrant Education Program:

COLLEGE ASSISTANCE MIGRANT PROGRAM (CAMP)

Purpose: CAMP provides funds to institutions of higher education or private nonprofit organizations to assist migrant and seasonal farmworkers enrolled as first-year undergraduates to make a successful transition from secondary to post-secondary education.

HIGH SCHOOL EQUIV-ALENCY PROGRAM (HEP)

Purpose: HEP funds projects at institutions of higher education or private nonprofit organizations that prepare migrant and seasonal farmworkers to obtain a High School Equivalency Certificate (GED) and assist them with their transition to a higher education institution, job training program or full-time employment.

INTER/INTRASTATE COORDINATION CONTRACTS (SECTION 1203)

Purpose: Inter/intrastate contracts are intended to improve the coordination of education and support services for migrant children nationwide:

The Migrant Student Record Transfer System (MSRTS) is a computerized system of transferring academic and health data on migrant children as they move between State and local education agencies.

Program Coordination Centers (PCCs) promote inter- and intrastate coordination. The PCCs, each of which is responsible for one of three migrant "streams," (the eastern, the central, and the western streams) respond to State requests for information,

coordination projects and
topical meetings.

The Migrant Stopover Site, in
Hope, Arkansas, provides
identification and recruitment
services for the Office of
Migrant Education (OME). In
addition, educational materials,
health brochures and counseling
services are provided.

The National Project for
Secondary Credit Exchange and
Accrual is working to resolve
the problems migrant students
face when they attempt to
transfer school credits from
State to State.

MIGRANT EDUCATION EVEN
START (MEES)

Purpose: The MEES program
funds family-centered education
projects that help migratory
parents become full partners in
the education of their children;
assist these children in reaching
their full potential as learners;
and provide literacy training for
parents by integrating early
childhood education and adult
education.

STATE BASIC GRANT
(SECTION 1201)

Purpose: The largest portion of
Migrant Education Program
funds is provided to State
education agencies to
implement instruction and
support programs for migrant
children and youth ages 3-21.

ENVIRONMENTAL
PROTECTION AGENCY

PESTICIDE PROGRAM

Purpose: The pesticide program
protects farmworkers, mixers,
loaders, applicators and other
workers exposed to pesticides.
It develops regulations to
protect agricultural workers
from pesticide hazards and
educational materials on
pesticide safety.

Contact:
James Boland, Chief
Occupational Safety Branch
Pesticide Programs
Environmental Protection
Agency
401 M Street, S.W.,
Room H7506C
Washington, DC 20460
Telephone: (703) 305-7666
FAX: (703) 305-5884

HEALTH AND HUMAN
SERVICES

COMMUNITY SERVICES
BLOCK GRANT PROGRAM

Purpose: The Community
Services Block Grant Program
provides funds to States for a
range of programs and activities
designed to alleviate the various
causes of poverty at the
community level.
A special set aside of $300,000 is
available to Historically Black
Colleges and Universities to

enable them to offer continuing education to migrant and seasonal farmworkers and to increase participant employment opportunities.

Contact:
Bobby Malone, Community Development Specialist
Migrant and Seasonal Farmworkers
Office of Community Services
U.S. Department of HHS
901 D Street, S.W., 5th Floor
Washington, DC 20447
Telephone: (202) 401-9343
FAX: (202) 401-4683/5718

MIGRANT HEAD START

Purpose: Migrant Head Start provides a comprehensive early childhood program for migrant preschool children, from birth to age 5.

Contact:
Frank Fuentes, Director
Migrant Head Start
U.S. Department of HHS
Switzer Bldg., Room 2226
330 C Street, S.W.
Washington, DC 20202
Telephone: (202) 205-8455
FAX: (202) 205-9721/8221

MIGRANT HEALTH PROGRAM

Purpose: The Migrant Health Program provides comprehensive primary health care services for migrant and seasonal farmworkers and their dependents.

Contact:
Antonio Duran, Director
Office of Migrant Health
Department of HHS
5600 Fishers Ln., Room 7A55
Rockville, MD 20857
Telephone: (301) 443-1153
FAX: (301) 443-4809

DEPARTMENT OF JUSTICE

SPECIAL AGRICULTURAL WORKER PROGRAM (SAW) and H-2A PROGRAM

Purpose: The Immigration Reform and Control Act of 1986 (IRCA) pertains to all employers of immigrant "special agricultural workers" engaged in seasonal agricultural work. The Immigration and Naturalization Service monitors the flow of such workers into the country. The H-2A program provides temporary, foreign farmworkers for employers who meet certification requirements. The program permits employers who meet requirements to import temporary farmworkers, provided no U.S. workers are available.

Contact:
Jack Rasmussen, Senior
Examiner
Immigration and Naturalization
Service
425 "I" Street, N.W., Room 7122
Washington, DC 20536
Telephone: (202) 514-5014
FAX: (202) 514-0198

DEPARTMENT OF LABOR

MIGRANT AND SEASONAL FARMWORKER PROGRAM (JTPA, SECTION 402)

Purpose: Provides training and support services, including job placement, for migrant and seasonal farmworkers to assist them in securing stable, year-round employment that provides income above the poverty line.

Contact:
Charles Kane, Branch Chief
Migrant and Seasonal
Farmworker Programs
200 Constitution Ave., N.W.,
Room N-4643
Washington, DC 20210
Telephone: (202) 219-5500
FAX: (202) 219-6338

MIGRANT AND SEASONAL FARMWORKER SERVICES (MSFW)

Purpose: MSFW provides funding for monitor advocates who provide advocacy services to domestic U.S. Migrant and Seasonal Farmworkers. These services include training, job referral and monitoring payments for required withholdings.

Contact:
Alicia Fernandez
National Monitor Advocate
U.S. Department of Labor
200 Constitution Ave., N.W.,
Room N-4456
Washington, DC 20003
Telephone: (202) 219-5174
FAX: (202) 219-6914

MIGRANT AND SEASONAL AGRICULTURAL WORKER PROTECTION ACT (MSPA) PROGRAM

Purpose: Program ensures that those who hire, employ, furnish and/or transport migrant and seasonal workers pay these workers according to fair labor standards, provide them with safe transportation, and offer them safe housing. Only employers that meet these criteria are registered and given permission to hire migrant and seasonal farm workers. The Act gives authority to investigate for non-registered employers and bring them into compliance with the provisions of the MSPA.

Contact:
Mrs. Corlis Sellers, Director
Division of Farm and Child
Labor and Polygraph Stds
200 Constitution Ave., N.W.,
Room S3510
Washington, DC 20210
Telephone: (202) 219-4670
FAX: (202) 219-5122

TEMPORARY ALIEN
AGRICULTURAL LABOR
PROTECTION PROGRAM

Purpose: Enforces the labor
protection standards for
temporary alien agricultural
labor allowed to enter and work
in the United States under the
H-2A Program.

TEMPORARY ALIEN
AGRICULTURAL LABOR
CERTIFICATION PROGRAM
(H-2A)

Purpose: Provides temporary,
foreign farmworkers for
employers who meet
certification requirements.
Permits employers who meet
requirements to import
temporary farmworkers,
provided no U.S. workers are
available.

Contact:
John Hancock, Manpower
Development Specialist
Division of Foreign Labor
Certification
200 Constitution Ave., N.W.,
Room N4466
Washington, DC 20003
Telephone: (202) 219-8660
FAX: (202) 219-6914

**NATIONAL NON-FEDERAL
ORGANIZATIONS
PROVIDING SERVICES TO
MIGRANT AND SEASONAL
FARMWORKERS AND
THEIR FAMILIES**

AMERICAN ASSOCIATION
OF RETIRED PERSONS

Purpose: The American
Association of Retired Persons
(AARP) published the booklet
"After the Harvest: The Plight
of the Older Farmworkers,"
which focuses on the need for
affordable housing for the
retired farmworker and
describes some existing projects
in this area.

Contact:
Leah Dobkin, AARP
Consumer Affairs
601 E Street, N.W.
Washington, DC 20049
Telephone: (202) 434-6030

CHILDREN'S DEFENSE FUND

Purpose: Provides advocacy for health care access, including advocacy for making Medicaid, child care and other Federal services available across State lines.

Contact:
Lourdes Rivera
Children's Defense Fund
Maternal and Child Health
Division
25 E Street, N.W.
Washington, DC 20001
Telephone: (202) 628-8787

THE COUNCIL OF CHIEF STATE SCHOOL OFFICERS

Purpose: The Council of Chief State School Officers sponsors the Interstate Migrant Education Council (see description below).

Contact:
Nancy Clark
Council of Chief State School
Officers
One Massachusetts Avenue,
N.W., Suite 700
Washington, DC 20001-1431
Telephone: (202) 408-5505

FARMWORKER JUSTICE FUND

Purpose: Litigation, public education, policy analysis and advocacy for migrant and seasonal farmworkers.

Contacts:
Michael Hancock, Executive
Director
Valerie A. Wilk, Health
Specialist
Farmworker Justice Fund
2001 S Street, N.W., Suite 210
Washington, DC 20009
Telephone: (202) 462-8192

HOUSING ASSISTANCE COUNCIL

Purpose: Provides training and technical assistance and loans for start-up of labor housing projects. Has a revolving loan fund to assist in land acquisition and technical assistance in rural development.

Contact:
Joe Belden
Housing Assistance Council
1025 Vermont Avenue, N.W.,
Room 606
Washington, DC 20002-4205
Telephone: (202) 842-8600

INTERSTATE MIGRANT EDUCATION COUNCIL (IMEC)

Purpose: Develops broad-based understanding among education, business and government decision makers in order to create an awareness of the unique needs of the migrant student population. Facilitates opportunities for interstate cooperation through sharing of model programs for migrant students. Identifies major barriers and develops alternative solutions for minimizing difficulties attributable to student mobility, intermittent attendance, limited English proficiency and other factors.

Contacts:
Nancy M. Clark, Administrative Assistant
John P. Perry, Project Consultant
Jim L. Gonzales, Policy Analyst
IMEC
One Massachusetts Ave., N.W., Suite 700
Washington, DC 20001-1431
Telephone: (202) 336-7078

MIGRANT LEGAL ACTION PROGRAM

Purpose: The program operates as a national legal service support center which provides representation to migrant and seasonal farm workers and provides technical assistance to migrant and other legal services field programs nationwide.

Contact:
Roger Rosenthal
Migrant Legal Action Program
2001 S Street, N.W., Suite 310
Washington, DC 20009
Telephone: (202) 462-7744

MIGRATION AND REFUGEE SERVICES/PASTORAL CARE OF MIGRANTS AND REFUGEES

Purpose: Pastoral Care of Migrants and Refugees (PCMR), a branch of the U.S. Catholic Conference, looks at issues surrounding the influx of migrants and refugees into the Catholic Church and how the Church should respond. An office within PCMR works directly with agricultural workers, providing advocacy and pastoral care.

Contact:
Father Richard Ryscavage, Executive Director, MRS
Father Peter Zendzian, Director, PCMR
U.S. Catholic Conference
3211 4th Street, N.E.
Washington, DC 20017
Telephone: (202) 541-3182

NATIONAL ASSOCIATION OF COMMUNITY HEALTH CENTERS

Purpose: This association represents the interests of Migrant Health Centers in Washington.

Contact:
Freda Mitchum, Associate
Director
National Association of
Community Health Centers
1330 New Hampshire Ave.,
N.W., Suite 122
Washington, DC 20036
Telephone: (202) 659-8008

NATIONAL COALITION OF HISPANIC AND HUMAN SERVICE ORGANIZATIONS

Purpose: To improve the health
and psycho-social well-being of
the nation's Hispanic population
through policy and research and
community-based research and
demonstration projects.
Migrants are reached through
community-based projects.

Contact:
Mary Thorngren
National Coalition of Hispanic
and Human Services
Organizations
1501 16th Street, N.W.
Washington, DC 20036
Telephone: (202) 387-5000

NATIONAL COUNCIL OF LA RAZA

Purpose: The National Council
of La Raza is the principal
constituency-based organization
serving Hispanics in such areas
of concern as employment
opportunities, discrimination,
anti-poverty and health
programs nationwide.

Contact:
Alejandro Becerra
Coordinator of Migrant
Activities
National Council of La Raza
810 First Street, N.W., Suite 300
Washington, DC 20002-4205
Telephone: (202) 289-1380

RURAL COMMUNITY ASSISTANCE PROGRAM

Purpose: A national network of
field-based, nonprofit
organizations working in small
rural communities – particularly
low-income and underserved
communities – to secure access
to safe drinking water and
sanitary waste water disposal
services, and to address solid
waste, ground-water and other
related issues. Handles the
Migrant Environmental Services
Assistance (MESA) Project.
Through MESA, RCAP has
assisted migrant health centers
to improve the living and
working conditions of MSFWs
with more than 100 projects in
25 States and Puerto Rico.

Contact:
Rose Holden
Rural Community Assistance
Program
602 King Street, Suite 402
Leesburg, VA 22075
Telephone: (703) 771-8636
DC Area: (703) 478-8652

Appendix C. Farm Labor Data

This book is about the endless quest of federally-funded programs that help needy migrant and seasonal farm workers (MSFWs). We have explained that when these assistance programs were begun in the mid-1960s, farm workers who found employment in the United States were different. There were more farm workers then, they were dispersed across more states and more farms, and most of them were U.S. citizens. Today there are fewer farm workers, they are concentrated on fewer farms in fewer states, and most are not U.S. citizens.

These broad strokes paint the picture of the farm labor market, but it is very hard to fill in the details because farm labor data are woefully inadequate. Most U.S. employment data is produced by the U.S. Department of Labor (DOL), which analyzes data collected every month to determine how many Americans are employed and unemployed and, for those who are employed, their hourly earnings and weeks worked. The DOL also has primary responsibility for operating the training programs that serve unemployed and disadvantaged workers. This means that one federal agency – DOL – produces all three of the major types of employment data: establishment or employer data, household data, and administrative data.

The DOL data system does not fully cover agriculture. For this reason, it is easy to understand why there are arguments about how many farm workers there are, who they are, and how needy they are. Most of the agricultural establishment or employer-reported data are collected by the USDA, most of the household or personal interview data come from the Commerce Department's Census Bureau or DOL, and the administrative data on farm workers who received services from assistance programs are generated by a panoply of federal and independent agencies, including the Education Department, HHS, DOL, and the Legal Services Corporation.

These data measure different subparts of the farm labor market with various levels of precision.[1] No federal agency or council pieces together this welter of data to generate a consensus picture of who farm workers are and where they work, helping to explain why the USDA reports that there are fewer than 200,000 migrant workers, while DOL reports 800,000 to 900,000, and MHS 1 to 2 million.

This appendix does not try to combine the various types of farm labor data to paint a picture of the number, characteristics, and distribution of farm workers. Instead, it reproduces the production, employment, and wage data available that: (1) can be arrayed by state or multi-state region, and (2) have been used or would be likely to be used to paint the farm labor picture that many service providers want.

There are three parts to this appendix. First, the major types of employment and wage data are described. Second, the concepts and methods used to generate some of the major data sources are explained. Third come tables that array the data. With these data, you too can be a farm labor analyst!

Types of Data

Employment and wage data record what happens in labor markets—the number of people hired and their characteristics; the wages they are paid and their fringe benefits; and how long they stay with a particular employer or in a certain industry or occupation. No single data source can give a complete picture of the people or the conditions in a particular labor market. Instead, a labor market can be imagined as a room of unknown size and shape, and each data source can be thought of as a window which provides a view into the room. The completeness of the data in representing the entire labor market is indicative of the size of the window, and the reliability of the data is analogous to the quality of the view through the window and into the room.

There are three major types of labor market data. Establishment or employer-reported data are obtained from employers. Most labor market data are obtained from employers because it is cheaper to survey a sample of or to take a census of the nation's 7 million employers than to interview a sample or census of the 140 million persons in the U.S. labor force sometime during each year. Establishment data usually describe jobs. They report, for example, the number of employees at work during a survey week or month; the average hourly earnings they were paid as well as the hours they worked and the value of any benefits employees they received; and the employees' duration of employment with this employer.

Household data are collected from individuals and households. These data include the personal characteristics of workers as well as data on spells of unemployment and movement between employers, industries, and occupations. Some household data do not change (race and sex), other data change in a predictable fashion (age), and some can be obtained only through repeated interviews (employment status).

Establishment and household data can be collected through censuses or sample surveys. A census obtains data from everyone; a sample from only a subset of the group. Data obtained from random samples of establishments or households can be expanded to indicate what a census would have reported within certain bounds. For example, a sample wage of $5 ± 25¢ means that 95 percent of the time a census (or another sample) would produce a wage of $4.75 to $5.25.

Administrative data provide a third source of farm labor data. Administrative data are collected for many reasons. Employers file quarterly reports with tax authorities in order to pay social security and other payroll taxes. Farm labor contractors and some other farm employers must provide information about themselves when they register and obtain licenses to operate. Farm worker service agencies record who they serve.

Administrative data can be censuses, as UI and FLC registration are supposed to be, or samples, as client data are because no service program serves all those who are eligible. Sample data from these three data sources are in Appendix Table 1.

Appendix Table 1. Farm Workers and Wages in Various Data Sources

Term	Sample Definition	Source	Sample Data Item	Issue
1. Farm and farmwork	All work done for wages on a place which sells farm products worth $1,000 or more	Census of Agriculture	In 1987, COA labor expenditures were $12.7 billion	- Includes wages paid to hired workers and FLCs as well as family members, clerical workers, and corporate officers
				- May miss some agriculture service wages; labor expenditures are more than hired farmworker wages
2. Farm worker	a. Person who did farmwork for cash wages or salary	CPS-HFWF	In 1983, there were 2.6 million hired farmworkers, including 9 percent migrants	Sampling procedure based on housing units and interviews conducted in December, so many (Hispanic) farmworkers may be missed
	b. All persons employed on farms for wages during a particular period	ES-202	ES-202 reported that 44,000 crop and livestock employers hired an average 616,000 workers in 1986	Includes all types of workers employed on farms; usually covers workers on the payroll for the payroll period that includes 12th of month

Term	Sample Definition	Source	Sample Data Item	Issue
	c. Paid workers doing agricultural work during survey week by the type of farmwork they did	QALS	During the week of July 7-13, 1991, there were 3.7 million persons employed on U.S. farms, including 1.1 million hired workers and 0.4 million ag service workers	This survey conducted since 1910, may underestimate seasonal farmworker employment and may not generate reliable regional data
3. Migrant farm worker	Crosses county lines and stays away from home overnight to do farmwork for wages	CPS-HFWF	159,000 migrants employed sometime during 1985	Based on a sample conducted in December of about 1,500 households which include at least one person who did farmwork during the past year.; about 6 percent or 94 households included a migrant farmworker
	Does 25 to 150 days of farmwork annually, obtains at least half of annual income from farmwork, and cannot return home at the end of a workday	ES-223	Local ES staff estimate MSFW employment each month; in 1982, an average annual 62,500 migrants; 1/3 in California, and 90 percent in 10 states	No standard methodology for collecting data
	The children aged 3 to 21 of farmworkers who cross school district lines to do farmwork	ME-MSRTS	In 1982, about 190,000 currently migrant students (FTE) were identified, and 216,000 formerly migrant students (FTE)	Recruiters determine the eligibility of children; school districts get funding for each child they enroll as a migrant

Establishment Data

There are two major nationwide establishment sources of farm labor data: the quinquennial Census of Agriculture (COA) and the Quarterly Agricultural Labor Survey (QALS). The COA is conducted by the Bureau of the Census in cooperation with USDA; the QALS is conducted by USDA.

Census of Agriculture

The Bureau of the Census conducts a number of economic censuses, such as the Census of Manufacturers and the Census of Construction Industries. The Census of Agriculture (COA) is the Bureau's economic census of U.S. farms and ranches. The Census of Agriculture has been conducted periodically since 1840. It is conducted in years that end in 2 and 7, with the questionnaire actually mailed to farmers the year after the reference year, so that the 1992 COA questionnaire was mailed early in 1993. COA data are published for the United States, each state, and for 3,100 counties.

The Census of Agriculture is mailed to a continuously updated list of farms maintained by the Bureau of the Census. The COA attempts to enumerate all "places" that sold or normally would have sold farm products worth at least $1,000 in the reference year; it did so in 1987 by mailing 4 million questionnaires to farms and potential farms, and the COA enumerated 2.1 million farms. Although undercounting, misreporting, and refusing to report are sometimes problems with the Census of Agriculture, the COA is considered the most comprehensive source of farm-related data, especially for states and counties.[2] COA data are often used to benchmark or adjust sample farm data collected in non-Census years, much as decennial Census of Population data are used to benchmark the monthly Current Population Survey.

The COA collects demographic and employment data on one operator per farm. The person completing the questionnaire determines who the operator is, and the operator may be the owner, a tenant, a paid manager, or a sharecropper. The COA collects data on the age and principal occupation of these farm operators, and the 1987 COA reported that the average age of farm operators was 52, that 94 percent of the farm operators were male, and that 45 percent of all farm operators were not primarily farmers. About 40 percent of all farm operators did 150 or more days of nonfarm work. Over 97 percent of all farm operators were white in 1987. The COA does not collect demographic or employment data on unpaid family workers.

Farm operator data are collected from all U.S. farms, but farm employment data are obtained from a sample of farms which is stratified so that almost all large farms are asked to complete a questionnaire and 1 in 6 small farms report employment data. The COA has consistently obtained data on the number of farms that hire labor and their payroll during the year preceding the COA. In some years, the COA has also included questions on employment, but "the utility of this information has been marred by an incredible variety of factors,"[3] including changing the employment reference week and the wording of the question. Employment questions were included in the 1978 and 1982 COA, dropped from the 1987 COA, and returned to the 1992 COA questionnaire.

The employment questions in 1982 were straightforward; they asked a sample of farmers whether they employed farm workers, whether any hired workers they employed were employed for more or less than 150 days on their farm, and their expenditures for hired labor, contract labor, and custom work.[4] In 1982, the COA reported that almost 4 million "seasonal" workers were employed less than 150 days on farms and about 1 million "regular" workers were employed for 150 days or more. This does not mean the COA reported that 4 million "seasonal" and 1 million "regular" workers were hired by the nation's farmers in 1982. It means there were 5 million jobs created on farms; a "seasonal" worker employed for 30 days by three different farmers is reported three times in the COA. COA employment data do not distinguish between family and nonfamily hired workers, so the "regular" and "seasonal" workers reported in the COA might include a son in partnership with a father and teenagers who are paid to work during the summer on their family's farm.

COA employment data can be cross-tabulated in a variety of ways. For example, the volume of employment can be reported by commodity: In 1982, there were 33,400 fruit, vegetable, and horticultural (FVH) specialty farms that employed workers for 150 days or more, and they employed 279,000 or 30 percent of all "regular" workers reported in the COA. About 64,000 FVH farms employed workers for less than 150 days in 1982, and they employed 1.2 million or 30 percent of all "seasonal" workers.

The Census of Agriculture dropped the workers-employed question from the 1987. In 1987, USDA asked employers to report their expenditures on "hired farm and ranch labor," including "hired workers, family members, hired managers, administrative and clerical employees, and salaried corporate officers." Labor expenditures include gross wages or salaries, employer-paid payroll taxes, the costs of any fringe benefits such as health insurance and any commissions paid to obtain workers.

COA labor expenditures are the only farm labor data that are collected on a consistent basis and available at the state and county level. Hours worked are not reported, simply expenditures, but if factors such as commissions and family pay are fairly uniform, then farmer-reported labor expenditures can be divided by an independent average hourly earnings data to estimate hours worked. For example, California farmers reported paying 31 percent of the labor expenditures of crop farms in the 1982 COA and Florida farmers paid almost 9 percent. The USDA's QALS (see below) reported that the average field worker[5] earned $4.69 in California in July 1982, and $3.90 in Florida. If these crop labor expenditures are divided by hourly earnings, California's share of total crop hours worked falls to 26 percent and Florida's rises to 14 percent.

In 1987, labor expenditures were $12.7 billion. Field and general crop farms had $3.1 billion in expenditures, or about one-fourth of the total, while FVH commodities accounted for $5.1 billion,[6] or 40 percent of total labor expenditures. Livestock farms accounted for $4.5 billion, or 35 percent of total labor expenditures. All states except four -- Alaska, Massachusetts, New Hampshire, and Rhode Island -- saw their farm labor expenditures rise between 1982 and 1987 and, in most states, FVH labor expenses rose at a faster rate than all labor expenditures.

Three states accounted for 5 percent or more of the nation's farm labor expenditures in 1987: California, Florida and Texas. California alone accounted for $3 billion or 24 percent of all farm labor expenditures and 32 percent of all crop labor expenditures. Florida farmers reported farm labor expenditures of about $1 billion in 1987, or eight percent of total expenditures. Texas farmers reported farm labor expenditures of $773 million or about six percent of the total. Florida labor expenditures were weighted toward FVH farms, while Texas labor expenditures were made mostly by livestock farms.

The COA permits the labor expenditures of farms in the various crop subcategories to be compared. Illinois farms accounted for 10 percent of the labor expenditures made by cash grain farms, and Arkansas and Nebraska farmers reported another seven percent each. The labor expenditures of cotton farms are similarly concentrated in a handful of states: California and Texas accounted for over half of the labor expenditures of cotton farms. North Carolina farmers accounted for nearly half of the $195 million spent on labor in by tobacco farms.

Single states accounted for half or more of the labor expenditures of farms in some types of crops, e.g. California accounted for 55 percent of the labor expenditures of fruit and nut farms. The labor expenditures of livestock farms were not as concentrated: California and Texas each accounted for about 9 percent of livestock labor expenditures, but over 30 states each accounted for 1 percent or more of livestock labor expenditures.

Quarterly Agricultural Labor Survey (QALS)

The QALS data series, published by USDA in *Farm Labor*, is one of the few farm labor data series used in regulatory proceedings: data on the hourly earnings of field and livestock workers are used by the U.S. Department of Labor to determine the Adverse Effect Wage Rate (AEWR), the wage rate that employers requesting temporary alien workers under the H-2A program must offer to both U.S. and alien workers to have their request for temporary H-2A alien workers certified or approved by DOL. QALS data are also used in many farm labor market studies to indicate, e.g., how farm worker employment has changed over time and how farm wages change relative to nonfarm wages.

USDA has conducted a farm employment survey since 1910. Between 1910 and 1974, USDA relied on "volunteer" farmers to report their farm employment and wages each month. In 1974, the survey was converted to a probability survey and conducted quarterly, hence QALS. Budget cuts in the early 1980s led to several years of limited data collection, but between 1984 and 1990, the QALS was restored to a quarterly survey. The survey was changed in 1991 to provide limited employment and wage data for 11 states (NY, PA, NC, FL, MI, WI, TX, NM, OR, WA, CA) each month and more complete data for 18 multi-state regions in January, April, July, and October.

The QALS sample is drawn from the list of farm employers maintained by USDA to conduct most of its periodic surveys, such as the number of livestock on farms and the acreages of various crops. The USDA list of sample farm employers is drawn from two sources or frames: (1) a list frame, or a list of farms known or likely to employ farm labor because of their size or major commodity, and (2) an area frame of about 16,000 land units in the United States. The area frame is used to supplement or correct for omissions in the list frame, and the master sample list is then corrected for duplication. For example, about 1,200 farms are interviewed each quarter in California; 1,000 are drawn from the list frame and 200 from the area frame.[7]

There is little specific information available on the QALS sampling procedures. The master list which is contacted, usually by telephone, to conduct the survey each quarter includes about 13,500 farms nationwide. This is a stratified sample, apparently drawn from employers who are pre-grouped by the number of workers employed or wages paid. About 50 percent of the sample is rotated each quarter so that over four quarterly surveys, about 27,000 U.S. farms might be contacted by NASS interviewers.

Only 50 to 70 percent of the employers contacted provide employment and wage data, i.e., 30 to 50 percent of the employers contacted did not hire workers during the survey week or they refused to cooperate. In an 1987 evaluation of the QALS, it was revealed that only 8,000 to 8,500 of the sample farms reported that they hired any labor, and that only 4,000 to 4,500 report they hired field and livestock workers during the survey week.

QALS defines farm workers to include all the people employed on crop (01) and livestock (02) farms except those who regularly do nonfarm work, such as those who do carpentry or domestic household work, or those do work that materially changes the form of the product, so that packing peaches into boxes on the farm is farm work but canning them is not.

The QALS survey obtains employment data on farm operators, unpaid workers, and hired workers. The operator is asked how many hours he and each of his partners worked during the survey week (the Sunday through Saturday that includes the 12th day of the month). Operators are also asked about family and other workers who did at least five hours of work during the survey week without pay. This unpaid group is heterogeneous: It includes children who help with farm chores if they receive an allowance but not a wage based on time worked, prisoners, and workers on Indian and religious farms. These questions generate employment and hours of work data, such as during the week of July 8 - 13, 1990, an estimated 1.6 million farm operators (including partners) averaged 41 hours of work on their farms, and 605,000 unpaid workers averaged 38 hours of work on farms during the survey week.

Hired workers include all people on the farm's payroll during the survey week, including paid family members, part-time workers, and hired managers. Hired workers are reported by their duration of employment on the responding farm (more or less than 150 days) and by the type of work they do (field, livestock, supervisor, and other such as bookkeeper). The third reporting dimension is the method by which they are paid (hourly, piece rate, or other such as salary). Workers doing a variety of tasks on the farm are classified by the type of farm that employs them, so that a general laborer on a crop farm is a field worker and a general laborer on a livestock farm is a livestock worker.

In July 1990, the responses from the sample of employers contacted were expanded to estimate that 650,000 hired workers would be employed more than 150 days and 456,000 would be employed less than 150 days on responding U.S. crop and livestock farms.

Farm Labor reports eight types of hourly earnings data for hired workers. Farm employers are asked to report the number of workers who are paid in a certain manner (say 10 field workers paid hourly wages), their total hours worked during the survey week (say 300), and their total gross earnings (say $1,500). Gross earnings are then divided by total hours to report that field workers who were paid hourly wages had average hourly earnings of $5.

This indirect method of computing hourly wages has the advantage of generating a single hourly earnings figure despite the diversity of agricultural wage systems. However, this can also be a weakness. Dividing gross wages by total hours "weights" reported hourly earnings by hours worked. For example, if a farm employs 10 irrigators at $4 hourly who each work 60 hours during the survey period, and 30 harvesters at $6 hourly who each work 25 hours, the average field worker wage for this farm is gross earnings ($2,400 + 4,500) divided by gross hours (600 + 750), or $6,900 ÷ 1350 hours = $5.11.[8] Even though there are three times more harvesters than irrigators, the average hourly earnings are near the midpoint of the two wage rates because the irrigators work more hours.

Gross earnings are divided by worker hours to generate average hourly wages for five types of workers (field, livestock, field and livestock combined, supervisory, and other) and three methods of pay (hourly, piece rate, and other). Gross earnings are what should appear on an employee's pay stub; that is, gross earnings exclude employer-paid payroll taxes and the costs of any fringe benefits provided to workers.

The QALS also collects data on fringe benefits. All hired workers must be assigned to one of six benefit categories, with benefits ranked in a preassigned order. Employers are asked how many workers receive or will receive--not necessarily during the survey week—(1) housing and meals, (2) housing only, (3) meals only, (4) cash bonuses, (5) other benefits such as health insurance, and (6) cash wages only. Workers can be assigned to only one of these six benefit categories so, e.g., many unionized California farm workers will appear in the fifth category. This benefit question gives no guidance on the proportion of hired workers receiving any specific benefit except the first.

The QALS asks operators about the value of their farm sales during the previous year and whether field crops, fruits and vegetables, or livestock and poultry contributed most of the farm's sales. Hired worker data for these cross-tabs were first published in 1988. In July 1990, for example, wages for U.S. field and livestock workers were reported to be $5; but $4.70 in field crops and $5.39 in fruits and vegetables. Larger farms, as measured by their farm sales, paid higher wages than smaller farms.

The QALS collects data on agricultural service workers, who are workers employed by firms that provide soil preparation services

(071), crop services such as combining grains or ginning cotton on the farm (072), veterinary (0741) and other livestock (0751) services, and farm labor and management services (076).[9] QALS asks farm employers in all states except California and Florida to report the type of agricultural service and the number of service workers employed on their farm during the survey week.

In California and Florida, QALS conducts a separate survey of agricultural service firms, which generates service worker data comparable to that obtained from farm employers. Florida and California state offices maintain a list of agricultural service firms, and interviewers try to obtain the name of each firm used on a sample farm to determine if it is on the statewide list. A sample of agricultural service firms is contacted during each quarterly survey. These service firms are asked the same employment and wages questions as farm operators, that is, they report workers, gross earnings, and total hours by type of worker and method of pay.

The QALS reports several hourly earnings figures. In 1990, farm workers earned an average $5.52 per hour, and this average ranged from $4.66 or 16 percent below the average in the Delta states of AR, LA, and MS to $8.61 or 56 percent above the average in HA. There appears to be an inverse relation between hourly wages and levels of employment: higher wages in HA are associated with relatively little employment, and lower wages in TX and OK are associated with relatively more employment.

Hourly earnings reported in the QALS increased about 127 percent between 1975 and 1990, but most of this increase occurred in the late 1970s. QALS hourly earnings rose only 50 percent between 1980 and 1990. California's earnings followed the U.S. pattern; they rose 118 percent between 1975 and 1990, and 40 percent between 1980 and 1990 (California's minimum wage rose 37 percent during the 1980s).

Year-to-year changes in hourly earnings are erratic, suggesting a high variance in the estimates. In Idaho, Montana, and Wyoming, hourly earnings rose 29 percent between 1980 and 1987, but the year-to-year changes during this period included a 12 percent drop one year, followed by a 34 percent increase, then a 10 percent drop, an 11 percent increase, a 23 percent increase, and finally an 11 percent drop. Such roller-coaster earnings more likely reflect problems with the QALS than the experience of workers.

Farm Labor reported that the average employment of hired workers was about 1 million in the mid-1980s, and that about two-thirds of the hired workers were regular or employed on the responding farm 150 days or more. Between 1975 and 1990, hired worker employment fell 32 percent throughout the United States, with the sharpest employment reductions concentrated in the southeastern states, in Appalachian states, in the Delta states, and in the cornbelt states of IA and MO.

Household Data

Household data are obtained from individuals, and households or individuals are the major source of data on the demographic characteristics of the population and work force, employment and earnings, and migration across occupations, industries, or areas. Some household information does not change over time, such as sex and race, some changes in a predictable fashion (age), and other household characteristics such as employment status and earnings can be quite variable from month-to-month and year-to-year, especially for seasonal farm workers. Thus, household questionnaires obtain employment and earnings information by asking about a particular reference week, month, or year.

The federal government conducts three household surveys which generate information on the characteristics and earnings of farm workers. The Census of Population (COP) asks a sample of households about the occupations of family members during the last week in March of the previous year. In the March 1980 COP, 1.33 million persons in the United States were classified as "farm workers and related occupations:" these workers were 70 percent White; 17 percent, Hispanic; 10 percent, Black; and 3 percent, Asian or other. In California, the census recorded 238,000 farm workers, almost 18 percent of the nation's total, and 53 percent of California's farm workers were Hispanic; 30 percent White; and 17 percent Asian or black. The "related occupations" in the census include nursery workers, nonfarm groundskeepers and gardeners, nonfarm animal caretakers, and graders and sorters of agricultural products.

The decennial Census of Population is a basic source of data on worker characteristics, but the occupational information refers only to persons who were doing farm work in March. Because this survey misses workers who were employed in months other than March, nearly two-thirds of U.S. farm workers may be missed.[10] Furthermore, the farm workers employed in March are different from workers employed only in the summer and fall. March farm workers tend to be older whites employed in regular jobs and dependent on farm work for almost all of their incomes.

A second federal source of farm worker characteristics and earnings data is a biennial household survey conducted for USDA through December 1987. The USDA's Hired Farm Working Force (HFWF) report is based on responses to questions attached to the regular Current

Population Survey of 60,000 households. In most years, about 1,500 households reported that someone in the household did farm work for wages in that year, including about 180 households in California. Responses from these sample households were then used to estimate the number, distribution, and characteristics of all farm workers.

The HFWF survey estimated that 2 to 2.5 million persons 14 and older did farm work for wages sometime during the year in 1980s. According to the HFWF, about three-quarters of these hired farm workers were White, one-eighth were Black or Hispanic, respectively. In the mid-1980s, farm workers earned an average $3,500 for 100 days of farm work, or an average daily wage of $35. Most farm workers do relatively little farm work; the one-sixth of all farm workers employed 250 days or more in agriculture contributed one-half of the work done by farm workers.

The Washington D.C.-based Association of Farmworker Opportunity programs (AFOP) has recommended that the COP-Questionnaire be modified (1) to add agriculture to the choice of industries in Q. 28c and (2) to separate farm and nonfarm wage and salary income in Q. 32a (Figure 1). Adding agriculture as an industry of employment and obtaining information on farm earnings could permit the COP to better distinguish farm workers from other workers and to generate more information

The basic CPS questionnaire obtains demographic and survey week employment data, and the HFWF supplementary questions ask for farm days worked and farm earnings for each month of the past year. All farm workers were asked if they left home and stayed away overnight to harvest crops or do other farm work for cash wages. A "yes" response to this question is followed by a question to determine if the overnight stay was in a different county. Thus, about 1,500 of the CPS households included at least one person who reported doing farm work for wages sometime during the year (2.5 percent), and about 120 of these 1,500 farm worker households (8 percent) included a person who stayed away overnight in a different county to do farm work in 1983.

USDA developed the definitions used to determine who was a migrant worker. The HFWF defined a migrant farm worker as a person who stays away from home overnight in a different county to do farm work for wages. Once a person is defined as a migrant farm worker in the HFWF, all of that person's farm work during the year is considered migratory farm work, so that even if the worker did one day of migratory farm work and 100 days of local farm work, the HFWF reports 101 days of migratory farm work. Migrant farm worker households are found in about 20 states, and the HFWF expands the sample migrants in these states to generate an estimate of the number of migrant farm workers in 10 multi-state regions. The small sample size

(120 migrant households) limits the reliability of the HFWF estimate of the number and the distribution of migrants in each region.

The CPS-HFWF survey data has been reported in a variety of ways over the past decade, but the data exhibit a remarkable consistency; about 2.5 to 2.7 million persons did farm work for wages sometime during the year, and 190,000 to 225,000 were migrants. However, the small migrant sample makes the regional distribution of migrants suspect; e.g. sharp changes in the number of migrants, such as the jump from 1,000 migrants in the Northeast in 1976 to 7,000 in 1983 or from 9000 in the Pacific states to 55,000, probably are as reflective of problems with the sample size and sampling procedures as they are of the distribution of migrants.

The CPS-HFWF data are the only national and historical data based on a statistical survey that generate farm worker characteristics data such as migrancy. In 1985, the HFWF reported that 159,000 of 2.5 million farm workers or 6 percent of all farm workers were migrants. Migrants ranged from 2 percent of all farm workers in the northeastern states to 15 percent in the southeastern states. Migrants were reported to be 72 percent White, 9 percent Black, and 19 percent Hispanic; 57 percent of all migrants did less than 75 days of farm work in 1985 and only 20 percent primarily did farm work for wages in 1985. Migrants had average total earnings of $5,700, versus $5,800 for non-migrants.

The number and characteristics of migrant farm workers located by the CPS has fluctuated since the mid-1970s. For example, the number of migrants was 115,000 in 1981 and 226,000 in 1983 and the percentage of White migrants rose from 45 percent in 1983 to 72 percent in 1985. The USDA's HFWF report notes that migrant worker estimates must be "interpreted cautiously because they are based on a small number of (sample) cases."

The HFWF survey was originally scheduled for December so that migrants could be interviewed at home. Changing migration patterns and the rise of illegal alien workers may be partially responsible for what are perceived to be "too few Hispanic" migrants in the HFWF.[11]

The HFWF reports that most migrant farm workers were interviewed in December in the state where they did farm work, suggesting that the distribution of the HFWF sample across states is a guide to both the residences and the workplaces of migrants.[12]

NAWS

The National Agricultural Worker Survey (NAWS) is a DOL-funded survey of farm workers whose purpose is to determine entries into and exits from the farm work force between 1979 and 1983. The

NAWS is a product of IRCA's agricultural provisions, and its future is uncertain. The characteristics of the workers interviewed in the NAWs were summarized in Chapter 5, where it was explained that, because the NAWS was designed to be a *national* sample of farm worker availability, no state-by-state NAWS data are available.

The NAWS interviewed a sample of about 2,500 new workers annually in 1990, 1991, and 1992, and re-interviewed about 2,000 workers one year after they were first interviewed. The sampling procedure has evolved, but the NAWS originally located workers to be interviewed by sampling their employers; the NAWS can be thought of as interviewing some of the workers employed by some of the employers on a QALS-type list of employers. The NAWS could not get access to the QALS list of farm employers, so it developed its own list of employers for the 72 counties and 25 states in which workers were interviewed.[13] It is for this reason that the NAWs is emerging as the primary source of data on the characteristics of workers employed in fruit and vegetable agriculture. However, it is not yet clear how NAWS data on farm worker characteristics such as migrancy could be used to determine the number of proportion of a subset of MSFWs in various regions and states. An initial attempt to distribute 2,115 interviewed workers by region indicated that the percentage of migrants varied by a factor of two. Migrants were disproportionately found in the southwest, southeast, and northwest; and relatively few were located in the northeast and midwest.

It has been proposed that NAWS data be used in a two-stage procedure to estimate the number, distribution and characteristics of farm workers eligible for migrant services. First, the regional number and distribution of employed workers would be taken from the QALS. Second, the NAWS would provide worker characteristics data, as well as data on unemployed farm workers. Third, to distribute migrant service funds within the multistate regions of the QALS, regional administrators could divide up the funds available to each region on the basis of detailed state and county data from the COA labor expenditure data.

Appendix Table 2. NAWS SAS Worker Characteristics by Region: 1990

Region	Workers Interviewed (%)	Foreign Born (%)	Poverty-level Income or Less (%)	Migrants (%)
Northeast	13	29	61	29
Southeast	19	81	73	48
Midwest	25	29	29	23
Southwest	26	91	56	24
Northwest	8	75	51	52
West	9	–	45	44

Source: Interviews with 2,115 SAS workers in 1990. Worker interviews are conducted in 72 counties in 25 states. The states in which regional interviews were conducted in 1990 were: NE(ME, NY, NJ, PA, WV, NC); SE(FL, GA, AL, LA); MW(IA, IN, IL, MI, WI, KY); SW(CA, AZ); NW(WA, OR); W(TX, OK, KS, CO, ID)

Administrative Data

Administrative data are generated when farmers pay unemployment insurance (UI) and social security taxes for their employees, when farm workers report their occupation on Internal Revenue Service tax forms, and when farm worker assistance programs obtain data from the clients they serve. Administrative data have not been used widely to chart trends in the farm labor market for several reasons. First, the coverage of farm workers under programs such as UI is incomplete in many states, so that administrative UI data paint only a partial picture of the farm labor market. Second, some of the administrative data collected on farm workers are of very inconsistent quality, such as the Department of Labor's ES-223 In-Season Farm Labor Reports. Finally, client data may be seen as self-serving and, since every farm worker assistance program asserts that it serves only a fraction of its target population, it is hard to adjust client data to estimate the target population to be served.

Administrative data have problems, but they also have advantages. First, since administrative data are already collected, they are relatively cheap to obtain and manipulate. Second, administrative data may provide more complete coverage than statistical sources; the UI data in California, for example, should be a "census" of all covered employers and their employees. Third, administrative data forms can be designed to collect precisely the data wanted, e.g. migrant clinic intake forms can be designed to collect the precise demographic and employment data desired.

This section reviews three administrative data sources. The first is the ES-202 or UI data which are compiled from the farm employers who are required to provide UI coverage to their employees. The second administrative data source is the ES-223 In-Seasonal Farm

Labor Reporting System; these are reports submitted by Employment Service staff in various states and published by the U.S. Department of Labor. Finally, the Migrant Student Record Transfer System (MSRTS) is described. This system reports data on the children of migrant farm workers.

ES-202 Employment and Wages Program or UI Data. The Unemployment Insurance (UI) program is a federal-state system enacted in 1938 to partially replace earnings lost due to unemployment. The federal government establishes minimum coverage levels (to determine which employers must cover their employees) and levies a tax on the first $7,000 earned by each covered worker to pay for the administration of the system. However, each state determines exactly what type of job loss constitutes unemployment and the level of benefits paid to unemployed workers.

The federal government mandated UI coverage of farm workers employed by large farm employers in 1978. Since then, farms that paid cash wages of $20,000 or more for farm labor in any calendar quarter in the current or preceding year, or which employed 10 or more workers on at least one day in each of 20 different weeks, must provide UI coverage for their employees. This "20-10" rule results in federal UI coverage only for farm workers employed by larger farm employers; coverage was estimated to be 40 percent of all farm workers in the mid-1970s, but the continuing concentration of farm production on fewer and larger means that in the 1980s about 70 to 80 percent of all farm workers are probably employed on farms that must provide UI coverage to their workers.

Federal UI law defines a farm employer as an entity that hires workers to raise or harvest agricultural or horticultural products on a farm; repair and maintain equipment on a farm; handle, process, or package commodities if over half of the raw commodity was produced by the employer; gin cotton; or do housework on a farm operated for profit. Seven states—California, Maine, Minnesota, Rhode Island, Virginia, Texas, and Washington—require additional agricultural employers to provide UI coverage for their employees. California has since 1978 required all employers who pay $100 or more in quarterly wages to provide UI coverage for their employees, and Texas has since 1987 required farm employers of three or more employees for 20 or more weeks, or who pay at least $6,250 in wages during any quarter, to cover their employees. Washington in 1990 eliminated virtually all farm employer UI exemptions. Different levels of coverage of smaller employers are reflected in state-by-state UI data.

Selected employment and wage data are forwarded by the states to the Employment and Training Administration in Washington, which publishes this data annually in *Employment and Wages*. Data are published for each state at the two-digit SIC level, that is, for crop, livestock, and agricultural service employers, and are available for more detailed commodity groupings, so that the number of employers (reporting units), average monthly employment, and total farm wages can be determined by commodity for each state. The UI system tabulates reporting units, not employers, so that a single employer with separate crop and livestock operations which each employ 50 or more workers would be two reporting units in UI data. Similarly, separate operations in two counties should generate a reporting unit in each county.

The UI system defines employees as all persons on the payroll, so that the paid managers of an agricultural corporation, supervisors, office personnel, professional staff, and field workers are all considered to be "farm workers". Farm employers report only cash wages; not payroll taxes, bonuses or the cost of fringe benefits. Averages computed from UI data must be interpreted carefully: "employers" may represent several sub-units of one large farm; reported employment increases with the length of the payroll period because of worker turnover; and wages represent pay to both professionals and to field workers.

UI data should be a "census" of covered employers and workers. This means that as more farm workers find employment on the larger farms which must provide UI coverage to their employees, the ES-202 data should become a more complete source of farm labor data. The UI system obtains basic employment and earnings data by commodity and county; the only data element missing to generate average hourly earnings is hours worked.

In 1990, over 150,000 farm and agricultural service employers reported an average annual employment of 1.4 million employees, and they reported total wages of more than $19 billion. As in COA data, crop farmers accounted for 29 percent of all reporting units, 38 percent of the employment, and 35 percent of wages paid. FVH farmers dominated crop UI data: 61 percent of all crop employers were FVH producers, and these growers represented 73 percent of crop employment and 72 percent of crop wages paid.

California was the dominant farm labor state, accounting for 23 percent of all U.S. farm employers who paid UI taxes, 31 percent of average annual employment, and 30 percent of UI wages paid. California's share of average annual employment was even greater in FVH (38 percent), farm agricultural services (56 percent), and farm labor contracting employment (66 percent).

Florida, Texas and Washington each had around 11,000 agricultural employers, and these states were second, third and fourth in average annual employment and wages paid by farm employers to UI authorities. But the combined average annual employment of these 3 states was less in 1990 than California's UI-covered farm employment.

California UI data confirm the widely-reported tendency of farmers to hire workers through FLCs. Average annual crop employment decreased 13 percent in California between 1980 and 1990, versus increases of 18 percent in Florida, 81 percent in Texas, and 122 percent in Washington; the latter two states expanded UI coverage to small growers during the 1980s. California increased its share of U.S. labor-intensive crop production during the decade; the decline in the state's crop employment was due to a shift toward hiring workers through agricultural service firms, whose employment rose 66 percent during the decade. Most of this surge in services employment was due to the expansion of farm labor contractor employment, which averaged 43,000 in 1980 and 75,000 in 1990 in California. The expansion of the FLC industry was almost entirely a California-specific phenomenon; by 1990, the state accounted for almost two-thirds of the nation's average FLC employment.

In-Season Farm Labor Reports (ES-223). The Department of Labor's In-Season Farm Labor Reports estimate the number of local and migratory workers employed during the week which includes the 15th of the month, although some states apparently use different survey weeks. The Employment and Training Administration (ETA) requires that local Employment Service offices in areas with 500 or more seasonal workers or any temporary alien H-2A workers to file monthly reports with ETA. ETA uses these data to operate its interstate job clearance system. For example, if West Virginia apple growers request temporary foreign workers, ETA can determine whether unemployed apple pickers are available in other states. There is apparently no sanction on states for not making such estimates; 13 states reported 0 migrants in 1982 (or did not file reports), 19 states reported 0 in 1985, and 15 states including Washington reported no migrants in 1987. ETA last published state-by-state 223 data in 1985, however, unpublished 12-page reports on MSFWs in each state are available.

The ES-223 reports divide farm workers into three groups: local seasonal workers (employed at least 25 days in agriculture and deriving at least half of their total incomes from farm work but not employed more than 150 days on any single farm); migrant workers (the subset of seasonal workers unable to return to their permanent residence at the end of the farm workday); and temporary H-2A workers. Agriculture is defined to include all crops (01); all livestock (02) except animal specialties (027); all agricultural services except veterinary services (074), animal specialty services (0752) such as farms boarding horses or dog kennels, farm labor contractors (0761), and landscape and horticultural services (078). This exclusion of the employees of labor contractors is curious. Full-time students working or traveling with their (farm worker) families are counted as seasonal workers, but not full-time students traveling in organized groups.

The ES-223 data define migrants as the subset of seasonal workers "whose farm work experience during the preceding 12 months" required an overnight stay away from home. ES-223 also defines migrant food processing worker as persons who were away from home overnight to work for up to 150 days with a food processing employer (SIC 201, 2033, 2035, or 2037) who packs, cans or freezes food products, and who earn at least half of their annual income in these three SICs.

State ES-223 reports list the crop activity, the stage of the work, and wages paid by county. A typical report in California (ED&R Report 881-A) indicated that during the week ending September 12, 1985, there were 22,800 workers employed to harvest 166,000 acres of raisin grapes in Fresno county; that the raisin harvest was 75 percent complete; and that wages averaged $3.35 to $3.50 per hour or 13¢ to 16¢ per tray. Data are reported by Agricultural Reporting Areas (ARA's), not necessarily counties, and ARAs are the approximately 219 (multi) county areas throughout the U.S. which employ 500 or more seasonal workers or H-2A workers.

ES-223 data are regularly criticized by researchers as inadequate. The criticisms of ES-223 data are many, but all relate to the fact that ES-223 data are not generated by a statistical survey and thus contain unknown errors. For example, the ARA's do not cover all agricultural areas in the United States, so ES-223 data would be incomplete even if each state filed reports. More serious is the absence of a formal survey or estimating procedure—ES personnel with multiple duties have said that they simply reported last year's numbers or estimated migrants by counting out-of-state license plates in rural areas or reading about farm workers in agricultural publications.

Vignettes from California and Texas give an idea of how ES-223 data are collected. California requests monthly data from the 31 Agribusiness Representatives (ABRs) who are responsible for farm labor activities in 42 of the state's 58 counties. These local ABRs have a handbook which outlines a procedure they can use to estimate the number of workers in their area on the basis of the acres of each crop and the hours of seasonal and other worker time needed to handle each acre. The quality of what is reported varies from ABR to ABR.

Texas followed a similar procedure of compiling of 50 local reports until 1986, when budget cuts caused the Texas Employment Commission (TEC) to stop making or reporting ES-223 data. When DOL insisted on getting ES-223 data from Texas, the TEC treated the whole state as one Agricultural Reporting Area and assigned one person to the job of satisfying DOL. That person reportedly adjusted the previous year's data on the basis of weather reports and farm labor and production articles.

In 1987, the ES-223 reported that there were 766,000 migrants employed for at least one month. About one-third were intrastate migrants and two-thirds were interstate. California had 37 percent of the reported total migrants, Florida had 14 percent, and Texas had 1 percent. However, the large and growing number of states with substantial agricultural production but no migrant or seasonal farm workers makes it very difficult to rely on ES-223 data.

Migrant Education. The Migrant Education is unique among assistance programs in its reliance on an in-house data source. Instead of adjusting the available farm labor data to estimate the number and distribution of the children eligible for ME services ME programs employ outreach workers who recruit eligible children, and then record them in the Migrant Student Record Transfer System (MSRTS). ME funds are then allocated to states, and to school districts within states, on the basis of the number of eligible children logged into the MSRTS. It should be emphasized that MSRTS data primarily reflect the number of ME-eligible children identified; they are not necessarily an indicator of the number of migrant farm workers, as occurs when most migrants are solo men. In such cases, there can be a great many migrant workers, but few children in the MSRTS. It is sometimes alleged that there are pockets of ME-eligible children who are not enrolled in the MSRTS but, given the financial incentive of school districts to locate eligible children, this may not be a significant problem.

Migratory students are entered into the MSRTS after they are enrolled by a local school district. Once enrolled, a migratory student remains "enrolled" in that school district for up to 13 months, even if the student "withdraws". The withdrawal days of migratory students are added to the days that migratory students actually attended school to determine total migrant FTE; withdrawal days are included in the count because Migrant Education funds are to be allocated on the basis of the number of days that migrant children reside in a state; not just the number of days that migrant children attend school in a state.

MSRTS data are available for 49 states and several U.S.-administered territories. All states except Mississippi have mandatory school attendance laws, but Migrant Education programs typically employ outreach workers to recruit migrant students instead of waiting for state education officials to find migrant children and compel them to attend school.

The MSRTS system produces a variety of data on the number and characteristics of the children logged into its system. These data are often reported on the basis of later or corrected tapes, making it important to use a consistent printout from MSRTS to examine trends over time. This section relies on the MSRTS printout of the "unique count of all migrant children reported as resident or enrolled during the 1980-81, 1985-86, and 1990-91 school years" of March 31, 1992.

The MSRTS printout assigns the migrant children reported to one of 6 categories. First, children are considered farming- or fishery-related according to whether the parent or guardian of the child did a qualifying farming or fishery activity. Second, within each of these qualifying activities, children are classified according to whether they moved with or to their parents within the past 12 months, and they are further subdivided according to whether this move was across state lines or across school district lines but within a state.

Between 1980-81 and 1990-91, the unique count rose 29 percent, from about 486,000 to 626,000. The growth in the formerly population was almost 3 times faster than the growth in the currently population: formerlies rose 42 percent to 330,000 in 1990-91, while currentlies rose 16 percent to 296,000 in 1990-91. The growth in these subcategories of migrant children reported by the MSRTS was faster in the second half of the decade. Between 1980-81 and 1985-86, the number of currentlies fell by 2 percent, while the number of formerlies rose by 13 percent. Between 1985-86 and 1990-91, the number of currentlies rose by 18 percent, while the number of formerlies jumped by 26 percent.

Migrant children reported to the MSRTS are concentrated in a handful of states. The top 20 states in 1990-91 each had about 1 percent or more of the migrant children reported to the MSRTS--30,000 or more currentlies and 33,000 or more formerlies in 1990-91-- included over 95 percent of all migrant children in the MSRTS.[14] ME children were concentrated in fewer than half of the states throughout the 1980s, but there have been significant shifts in state rankings for formerlies.

The top 5 ME states, which each had 5 percent or more of the currentlies and formerlies throughout the 1980s, saw their share of all currentlies drop from 76 to 70 percent, and their share of formerlies rise from 62 to 70 percent. The major reason for the drop in currentlies was that Texas reported about 10,000 fewer intrastate currentlies in 1990-91 than in 1980-81. In 1990-91, California included about 25 percent of the MSRTS currentlies, Texas 20 percent, Florida 12 percent, and Michigan and Washington 5 percent each.

There were more changes in state shares of formerlies during the 1980s. Texas's share of all MSRTS-reported formerlies fell almost in half, from 30 to 17 percent, and the drop was especially significant during the second half of the 1980s. California, on the other hand, almost tripled its share of all formerlies, from 15 percent in 1980-81, its share doubled to 30 percent in 1985-86, and then rose to 39 percent in 1990-91. Florida's share remained at 7 to 8 percent, while Puerto Rico and Louisiana, states which included almost 10 percent of the formerly migrant children in the MSRTS in 1980-81, had only a 5 percent share in 1990-91.

MSRTS is a system which provides detailed data on the children of current and former migrants, and thus should be able to generate data such as family size and migration patterns. However, the migrant characteristics in the MSRTS reflect the efforts of recruiters to enroll migrant children; if recruiters systematically enroll only children from families with certain characteristics, then the MSRTS will reflect the characteristics of just some migrants. More importantly, MSRTS data reflect the characteristics of migrant families, not solo male or skilled migrants. Thus, if particular groups of migrants are more likely to migrate as families (Texas Hispanic families versus single Florida Blacks), then MSRTS migrant data give distorted picture of the overall migrant work force.

Appendix Table 3. Farm Labor Expenditures Reported in the 1987 Census of Agriculture

State	All Farms Expend ($1,000)	% of U.S.	% Change 1982-87	Crop Farms Expend ($1,000)	% of U.S.	% Change 1982-87	FVH* Expend ($1,000)	% of U.S.	% Change 1982-87	Livestock Farms Expend ($1,000)	% of U.S.	% Change 1982-87
Alabama	143,370	1	52	60,781	1	67	30,205	1	72	82,589	2	43
Alaska	1,930	0	-4	1,768	0		1,260	0		162	0	-92
Arizona	248,041	2	32	193,893	2	38	119,067	2	45	54,148	1	14
Arkansas	225,122	2	30	119,632	1	-1	5,831	0	22	105,490	2	105
California	2,998,581	24	34	2,587,960	32	54	2,206,905	43	40	410,621	9	-26
Colorado	200,174	2	43	88,675	1	83	42,044	1	36	111,499	2	21
Connecticut	58,811	0	29	49,556	1	393	42,093	1	75	9,255	0	-74
Delaware	25,059	0	31	10,920	0	35	6,931	0	55	14,139	0	27
Florida	1,002,022	8	47	894,382	11	89	749,963	15	54	107,640	2	-49
Georgia	243,197	2	49	132,564	2	47	56,696	1	123	110,633	2	50
Hawaii	192,674	2	29	180,775	2	41	82,113	2	54	11,899	0	-45
Idaho	208,172	2	25	130,513	2	30	11,991	0	33	77,659	2	18
Illinois	291,228	2	25	188,346	2	40	59,606	1	51	102,882	2	3
Indiana	212,497	2	32	105,834	1	52	30,412	1	260	106,663	2	17
Iowa	303,545	2	32	109,756	1	21	11,826	0	-42	193,789	4	39
Kansas	242,526	2	50	77,295	1	7	8,719	0	49	165,231	4	85
Kentucky	180,267	1	1	73,193	1	-25	9,837	0	98	107,074	2	33
Lousiana	130,918	1	17	103,108	1	24	10,329	0	61	27,810	1	-3
Maine	56,480	0	17	31,143	0	24	14,029	0	27	25,337	1	9
Maryland	91,322	1	25	41,894	1	78	28,480	1	58	49,428	1	0
Massachusetts	43,206	0	-5	24,940	0	49	22,822	0	-8	18,266	0	-37
Michigan	270,179	2	35	178,264	2	80	137,250	3	46	91,915	2	-10
Minnesota	265,812	2	23	121,491	1	37	20,166	0	13	144,321	3	12
Mississippi	155,749	1	10	100,135	1	0	5,248	0	52	55,614	1	33
Missouri	185,226	1	23	87,151	1	35	19,683	0	51	98,075	2	14
Montana	115,687	1	28	47,020	1	15	2,688	0	57	68,667	2	38

223

State	Total											
Nebraska	281,729	2	61	93,374	1	35	2,448	0	15	188,355	4	77
Nevada	34,846	0	62	11,994	0	61	9	0	-98	22,852	1	62
New Hampshire	7,093	0	-50	862	0	-80	504	0	-92	6,231	0	-37
New Jersey	102,440	1	38	88,821	1	97	83,538	2	45	13,619	0	-53
New Mexico	118,631	1	69	53,388	1	81	32,456	1	76	65,243	1	61
New York	298,347	2	15	147,201	2	62	124,108	2	30	151,146	3	-10
North Carolina	329,127	3	24	193,099	2	5	42,183	1	100	136,028	3	65
North Dakota	101,079	1	25	74,556	1	18	1,094	0	26	26,523	1	51
Ohio	236,791	2	35	137,224	2	111	87,719	2	53	99,567	2	-10
Oklahoma	150,098	1	36	53,103	1	46	20,469	0	41	96,995	2	31
Oregon	286,595	2	48	224,710	3	108	151,971	3	72	61,885	1	-28
Pennsylvania	320,269	3	35	179,848	2	303	163,392	3	36	140,421	3	-27
Rhode Island	1,482	0	-74	967	0	-3	635	0	-86	515	0	-89
South Carolina	108,301	1	21	75,206	1	41	38,869	1	242	33,095	1	-8
South Dakota	104,829	1	48	29,569	1	36	1,888	0	137	75,260	2	54
Tennessee	135,418	1	15	77,192	1	41	38,033	1	92	58,226	1	-7
Texas	773,390	6	36	370,423	5	35	140,272	3	43	402,967	9	37
Utah	59,221	0	30	17,753	0	114	10,906	0	79	41,468	1	12
Vermont	32,159	0	10	4,119	0	81	4,119	0	86	28,040	1	4
Virginia	149,417	1	9	63,197	1	8	43,132	1	46	86,220	2	9
Washington	468,973	4	35	372,923	5	42	254,243	5	41	96,050	2	15
West Virginia	27,029	0	19	12,209	0	23	10,957	0	7	14,820	0	16
Wisconsin	329,160	3	14	93,220	1	51	50,888	1	55	235,940	5	4
Wyoming	60,404	0	35	12,110	0	40	205	0	13	48,294	1	34
United States	12,709,221	100	33	8,204,859	100	52	5,091,914	100	47	4,504,362	100	9

Source: Bureau of the Census, Census of Agriculture, 1987.

Labor expenditures include gross wages or salaries paid to hired workers and supervisors, bonuses, social security and other payroll taxes, and expenditures for fringe benefits. Labor expenditures include contract labor expenses.

*FVH includes labor expenditures in Vegetable, Fruit and Tree Nut and other Horticultural operations such as Nursery or Strawberries.

Appendix Table 4. Top Ten States in Farm Labor Expenditures Reported in the 1987 Census of Agriculture

State	All Farms		Crop Farms		FVH		Livestock Farms	
	Expend ($1,000)	% of U.S.	Expend ($1,000)	% of U.S.	Expend ($1,000)	% of U.S.	Expend ($1,000)	% of U.S.
California	2,998,581	23.6	2,587,960	31.5	2,206,905	43.3	410,621	9.1
Florida	1,002,022	7.9	894,382	10.9	749,963	14.7	402,967	8.9
Texas	773,390	6.1	372,923	4.5	254,243	5.0	235,940	5.2
Washington	468,973	3.7	370,423	4.5	163,392	3.2	193,789	4.3
Wisconsin	329,160	2.6	224,710	2.7	151,971	3.0	188,355	4.2
N. Carolina	329,127	2.6	193,893	2.4	140,272	2.8	165,231	3.7
Pennsylvania	320,269	2.5	193,099	2.4	137,250	2.7	151,146	3.4
Iowa	303,545	2.4	188,346	2.3	124,108	2.4	144,321	3.2
New York	298,347	2.3	180,775	2.2	119,067	2.3	140,421	3.1
Illinois	291,228	2.3	179,848	2.2	87,719	1.7	136,028	3.0
Top Ten States	7,114,642	56.0	5,386,359	65.6	4,134,890	81.2	2,168,819	48.1
United States	12,709,221	100.0	8,204,859	100.0	5,091,914	100.0	4,504,362	100.0

Source: Bureau of the Census, Census of Agriculture, 1987.

Labor expenditures include gross wages or salaries paid to hired and contract workers and supervisors, bonuses, social security and other payroll taxes, and expenditures for fringe benefits.
Crop Farms include all farms except livestock farms.
FVH includes labor expenditures in Vegetable, Fruit and Tree Nut and other Horticultural operations such as Nursery, Mushrooms, or Strawberries.

Appendix Table 5(A). Farm Labor Expenses Reported in the 1987 COA

State	Cash grains* Expend ($1,000)	% of U.S.	Fruit and Tree Nuts Expend ($1,000)	% of U.S.	Vegetables Expend ($1,000)	% of U.S.	Hort. Specialties Expend ($1,000)	% of U.S.	Other Crops Expend ($1,000)	% of U.S.
Alabama	4,458	0	1397	0	2198	0	26610	2	13,852	1
Alaska	0	0	0	0	0	0	1,260	0	508	0
Arizona	215	0	24,794	1	76,295	6	17,978	1	16,453	1
Arkansas	83,971	7	1,785	0	1,460	0	2,586	0	1,827	0
California	37,966	3	1,157,048	55	568,315	46	481,542	27	191,282	16
Colorado	25,660	2	4,701	0	15,125	1	22,218	1	20,971	2
Connecticut	0	0	2,947	0	1,111	0	38,035	2	602	0
Delaware	0	0	0	0	3,429	0	3,502	0	3,989	0
Florida	995	0	280,305	13	224,690	18	244,968	14	139,812	12
Georgia	7,970	1	16,391	1	11,199	1	29,106	2	52,581	4
Hawaii	0	0	60,900	3	5,720	0	15,493	1	98,662	8
Idaho	22,417	2	4,851	0	3,900	0	3,240	0	96,105	8
Illinois	124,455	10	4,667	0	9,538	1	45,401	3	4,285	0
Indiana	70,464	6	3,660	0	6,200	1	20,552	1	4,241	0
Iowa	95,541	8	823	0	746	1	10,257	1	2,389	0
Kansas	67,263	6	395	0	0	0	8,324	0	1,293	0
Kentucky	20,554	2	408	0	1,298	0	8,131	0	8,829	1
Louisiana	28,176	2	0	0	1,017	0	9,312	1	35,273	3
Maine	34	0	10,385	0	1,457	0	2,187	0	17,080	1
Maryland	9,008	1	3,367	0	4,110	0	21,003	1	2,998	0
Massachusetts	0	0	22,822	1	0	0	0	0	2,118	0
Michigan	19,894	2	44,685	2	29,831	2	62,734	4	21,120	2
Minnesota	61,928	5	1,270	0	4,188	0	14,708	1	39,397	3
Mississippi	25,833	2	468	0	342	0	4,438	0	3,790	0
Missouri	53,097	4	4,044	0	935	0	14,704	1	4,984	0
Montana	34,693	3	771	0	10	0	1,907	0	9,639	1
Nebraska	81,443	7	144	0	100	0	2,204	0	9,483	1
Nevada	0	0	0	0	0	0	0	0	11,985	1
New Hampshire	0	0	9	0	504	0	0	0	358	0
New Jersey	1,438	0	19,754	1	24,792	2	38,992	2	3,845	0

State	Cash grains* Expend ($1,000)	% of U.S.	Fruit and Tree Nuts Expend ($1,000)	% of U.S.	Vegetables Expend ($1,000)	% of U.S.	Hort. Specialties Expend ($1,000)	% of U.S.	Other Crops Expend ($1,000)	% of U.S.
New Mexico	2,260	0	11,043	1	15,900	1	5,513	0	14,423	1
New York	5,825	0	41,101	2	35,412	3	47,595	3	17,268	1
North Carolina	14,350	1	6,976	0	7,711	1	27,496	2	41,194	3
North Dakota	49,300	4	0	0	0	0	1,094	0	24,162	2
Ohio	40,439	3	6,575	0	18,813	2	62,331	4	8,236	1
Oklahoma	14,085	1	374	0	999	0	19,096	1	11,379	1
Oregon	13,126	1	60,624	3	28,982	2	62,365	4	59,613	5
Pennsylvania	6,673	1	24,838	1	7,154	1	131,400	7	9,624	1
Rhode Island	0	0	0	0	207	0	428	0	332	0
South Carolina	7,310	1	16,546	1	8,128	1	14,195	1	6,965	1
South Dakota	23,390	2	0	0	75	0	1,813	0	4,291	0
Tennessee	11,938	1	729	0	4,189	0	33,115	2	4,991	0
Texas	61,588	5	13,561	1	51,174	4	75,537	4	43,542	4
Utah	1,730	0	2,715	0	1,488	0	6,703	0	5,117	0
Vermont	0	0	2,512	0	83	0	1,524	0	0	0
Virginia	5,095	0	12,882	1	8,108	1	22,142	1	13,883	1
Washington	40,277	3	195,077	9	27,232	2	31,934	2	78,403	6
West Virginia	402	0	7,496	0	26	0	3,435	0	638	0
Wisconsin	14,236	1	20,248	1	14,096	1	16,544	1	27,745	2
Wyoming	2,112	0	0	0	0	0	205	0	9,793	1
United States	1,193,992	100	2,101,862	100	1,235,083	100	1,754,969	100	1,209,168	100

Source: Bureau of the Census, Census of Agriculture, 1982.

Labor expenditures include gross wages or salaries paid to hired workers and supervisors, bonuses, social security and other payroll taxes, and expenditures for fringe benefits.
Labor expenditures include contract labor expenses.

*Cotton labor expenditures were $515 million with 29 in California, 24 in Texas, 13 in Mississippi, 11 in Arizona and 6 in Louisiana; these 5 states accounted for 83 of total expenditures.

Tobacco labor expenditures were $195 million with 48 in North Carolina, 17 in Kentucky, and 10 in South Carolina.

Appendix Table 5(B). Farm Labor Expenses Reported in the 1987 COA

State	Beef, hog and sheep		Dairy		Poultry and egg		Other livestock	
	Expend ($1,000)	% of U.S.	Expend ($1,000)	% of U.S.	Expend ($1,000)	% of U.S.	Expend ($1,000)	% of U.S.
Alabama	22,337	1	7,375	1	49,838	6	3,039	1
Alaska	107	0	0	0	0	0	55	0
Arizona	32,464	2	18,487	1	717	0	2,480	1
Arkansas	21,999	1	4,779	0	72,501	9	6,211	2
California	81,984	4	193,386	15	114,046	14	21,205	7
Colorado	84,454	4	16,007	1	8,506	1	2,532	1
Connecticut	516	0	8,739	1	0	0	0	0
Delaware	1,357	0	1,540	0	11,169	1	73	0
Florida	32,667	2	41,198	3	14,324	2	19,451	6
Georgia	23,418	1	16,177	1	67,008	9	4,030	1
Hawaii	7,367	0	0	0	2,527	0	2,005	1
Idaho	42,613	2	22,453	2	2,438	0	10,155	3
Illinois	73,347	4	16,371	1	5,744	1	7,420	2
Indiana	56,589	3	14,704	1	28,868	4	6,502	2
Iowa	161,132	8	15,096	1	12,154	2	5,407	2
Kansas	152,051	7	8,452	1	1,898	0	2,830	1
Kentucky	35,776	2	21,587	2	2,258	0	47,453	15
Louisiana	9,843	0	8,675	1	6,728	1	2,564	1
Maine	630	0	10,038	1	11,976	2	2,693	1
Maryland	5,428	0	18,864	1	20,533	3	4,603	1
Massachusetts	971	0	7,273	1	2,743	0	7,279	2
Michigan	23,307	1	51,184	4	10,005	1	7,419	2
Minnesota	54,905	3	49,013	4	29,278	4	11,125	3
Mississippi	15,051	1	7,760	1	20,068	3	12,735	4
Missouri	62,132	3	17,944	1	14,520	2	3,479	1
Montana	60,353	3	3,541	0	1,074	0	3,699	1
Nebraska	173,883	8	6,805	1	3,165	0	4,502	1
Nevada	18,798	1	3,826	0	0	0	228	0

State	Beef, hog and sheep Expend ($1,000)	% of U.S.	Dairy Expend ($1,000)	% of U.S.	Poultry and egg Expend ($1,000)	% of U.S.	Other livestock Expend ($1,000)	% of U.S.
New Hampshire	271	0	5,960	0	0	0	0	0
New Jersey	1,062	0	6,712	1	2,443	0	3,402	1
New Mexico	46,335	2	14,038	1	3,387	0	1,483	0
New York	6,748	0	119,936	9	9,299	1	15,163	5
North Carolina	47,324	2	22,732	2	62,658	8	3,314	1
North Dakota	17,871	1	3,807	0	464	0	4,381	1
Ohio	32,357	2	38,264	3	17,571	2	11,375	4
Oklahoma	74,650	4	8,481	1	9,811	1	4,053	1
Oregon	32,739	2	18,831	1	6,919	1	3,396	1
Pennsylvania	21,699	1	72,875	6	33,360	4	12,487	4
Rhode Island	47	0	468	0	0	0	0	0
South Carolina	9,032	0	9,446	1	12,901	2	1,716	1
South Dakota	61,496	3	6,396	0	1,136	0	6,232	2
Tennessee	25,577	1	23,291	2	6,064	1	3,294	1
Texas	302,579	14	49,324	4	40,702	5	10,362	3
Utah	19,170	1	13,585	1	3,491	0	5,222	2
Vermont	1,129	0	25,637	2	587	0	687	0
Virginia	27,280	1	31,455	2	21,227	3	6,258	2
Washington	29,075	1	47,507	4	13,574	2	5,894	2
West Virginia	6,078	0	4,442	0	3,490	0	810	0
Wisconsin	30,517	1	177,040	14	10,225	1	18,158	6
Wyoming	46,199	2	1,334	0	10	0	751	0
United States	2094767	100	1,300,093	100	786,910	100	322,592	100

Source: Bureau of the Census, Census of Agriculture, 1982.

Labor expenditures include gross wages or salaries paid to hired and contract workers and supervisors, bonuses, social security and other payroll taxes, and expenditures for fringe benefits.
Other livestock includes all animals and animal products raised for sale and not included in the other categories.

Appendix Table 6. Agricultural Services Employment and Wages Covered by Unemployment Insurance in 1990

State	Employers Number	Employers % of U.S.	Employment Avg. Annual	Employment % of U.S.	Wages ($millions)	Wages % of U.S.	Percent Change 1980-90 Employers	Percent Change 1980-90 Employment	Percent Change 1980-90 Wages
Alabama	1,126	1.2	7,197	1.0	90.5	0.9	71	112	236
Alaska	104	0.1	442	0.1	6.7	0.1	137	205	219
Arizona	1,898	2.0	19,199	2.6	225.6	2.2	94	103	181
Arkansas	813	0.9	4,537	0.6	61.2	0.6	31	52	117
California	12,699	13.7	208,542	28.6	2,708.0	26.3	48	66	154
Colorado	1,494	1.6	9,413	1.3	133.4	1.3	64	67	151
Connecticut	2,004	2.2	8,243	1.1	157.8	1.5	81	88	245
Delaware	291	0.3	1,552	0.2	23.8	0.2	99	83	198
Florida	7,577	8.2	70,638	9.7	902.3	8.8	86	31	119
Georgia	2,102	2.3	14,564	2.0	207.6	2.0	104	157	300
Hawaii	310	0.3	3,025	0.4	55.9	0.5	52	73	229
Idaho	524	0.6	4,615	0.6	59.8	0.6	124	304	504
Illinois	3,226	3.5	25,325	3.5	427.8	4.2	87	129	226
Indiana	1,741	1.9	10,662	1.5	142.0	1.4	86	114	203
Iowa	1,196	1.3	6,686	0.9	84.7	0.8	23	70	117
Kansas	1,014	1.1	5,380	0.7	72.4	0.7	48	88	165
Kentucky	1,174	1.3	8,824	1.2	127.0	1.2	104	124	242
Louisiana	1,028	1.1	6,386	0.9	83.6	0.8	47	51	105
Maine	495	0.5	2,695	0.4	37.6	0.4	89	128	276
Maryland	2,011	2.2	14,044	1.9	236.8	2.3	106	124	291
Massachusetts	2,781	3.0	11,957	1.6	228.3	2.2	142	83	224
Michigan	2,774	3.0	16,930	2.3	271.0	2.6	103	98	189
Minnesota	1,512	1.6	8,804	1.2	141.2	1.4	46	40	126
Mississippi	874	0.9	4,371	0.6	54.0	0.5	54	35	105
Missouri	1,812	2.0	10,404	1.4	140.5	1.4	62	78	147
Montana	344	0.4	1,025	0.1	11.8	0.1	55	66	131
Nebraska	799	0.9	3,954	0.5	51.1	0.5	81	79	152

Nevada	493	0.5	3,394	0.5	51.6	0.5	116	210	282
New Hampshire	556	0.6	2,526	0.3	37.3	0.4	127	182	355
New Jersey	3,606	3.9	15,881	2.2	300.8	2.9	91	82	242
New Mexico	491	0.5	5,971	0.8	41.1	0.4	36	194	187
New York	5,610	6.0	25,646	3.5	477.4	4.6	47	67	195
North Carolina	2,279	2.5	15,388	2.1	211.9	2.1	108	119	279
North Dakota	264	0.3	974	0.1	13.9	0.1	34	51	121
Ohio	3,600	3.9	22,177	3.0	335.9	3.3	50	99	179
Oklahoma	964	1.0	4,924	0.7	63.0	0.6	74	45	106
Oregon	1,228	1.3	8,340	1.1	110.9	1.1	36	129	221
Pennsylvania	3,760	4.1	22,214	3.1	371.0	3.6	62	80	202
Rhode Island	506	0.5	1,847	0.3	31.6	0.3	64	94	251
South Carolina	1,173	1.3	7,247	1.0	97.8	1.0	91	100	242
South Dakota	327	0.4	969	0.1	13.3	0.1	75	28	68
Tennessee	1,283	1.4	8,720	1.2	111.6	1.1	79	76	187
Texas	5,616	6.1	43,964	6.0	554.9	5.4	64	54	112
Utah	412	0.4	2,337	0.3	26.5	0.3	66	109	157
Vermont	331	0.4	1,614	0.2	25.7	0.2	107	86	213
Virginia	2,046	2.2	15,644	2.1	240.7	2.3	108	155	346
Washington	2,465	2.7	15,234	2.1	198.1	1.9	58	89	150
West Virginia	308	0.3	2,247	0.3	28.4	0.3	39	81	193
Wisconsin	1,447	1.6	10,647	1.5	186.9	1.8	38	78	162
Wyoming	211	0.2	680	0.1	7.3	0.1	91	63	115
United States	**92,741**	**100.0**	**728,247**	**100.0**	**10,285.6**	**100.0**	**69**	**75**	**172**

Source: ES-202 Emplyment and Wages Program; Bureau of Labor Statistics, U.S. Department of Labor.

Most states have Federal Unemployment Tax Act coverage of farm employers; farm employers who pay quarterly cash wages of $20,000 or more or employ 10 or more workers for at least one day in each of 20 different weeks must cover their farmworkers (20-10 rule). CA, ME, MN, RI, TX, and WA have more inclusive farmworker coverage.

Percent changes may be exagerrated in states, including WA and TX, that extended coverage beyond the Federal minimum during the 1980s.

Employers are farms or their subunits which employ 50 or more workers.

Avg. Ann. Employment is the employer-reported count of all workers on the payroll for the pay period which includes the twelfth day of the month summed over 12 months and divided by 12

Appendix Table 7. Agricultural* Employment and Wages Covered by Unemployment Insurance in 1990

State	Employers Number	Employers % of U.S.	Employment Avg. Annual	Employment % of U.S.	Wages ($millions)	Wages % of U.S.	Percent Change 1980-90 Employers	Percent Change 1980-90 Employment	Percent Change 1980-90 Wages
Alabama	1,415	0.9	13,628	1.0	185.1	0.9	65	55	145
Alaska	116	0.1	565	0.0	8.5	0.0	134	152	193
Arizona	2,722	1.8	33,504	2.3	423.5	2.1	76	39	87
Arkansas	1,461	1.0	12,710	0.9	182.4	0.9	26	20	77
California	34,576	22.6	437,507	30.6	5,899.2	29.8	8	13	74
Colorado	2,125	1.4	17,733	1.2	260.6	1.3	65	54	131
Connecticut	2,212	1.4	13,573	0.9	251.1	1.3	75	34	193
Delaware	363	0.2	3,060	0.2	52.7	0.3	80	43	151
Florida	10,601	6.9	141,845	9.9	1,838.4	9.3	79	23	101
Georgia	2,672	1.7	25,746	1.8	356.6	1.8	88	73	194
Hawaii	548	0.4	11,895	0.8	233.0	1.2	64	6	61
Idaho	1,206	0.8	15,339	1.1	202.1	1.0	139	161	286
Illinois	4,009	2.6	36,537	2.6	607.3	3.1	85	93	184
Indiana	2,304	1.5	20,482	1.4	289.3	1.5	81	72	145
Iowa	1,703	1.1	11,510	0.8	159.6	0.8	33	26	60
Kansas	1,541	1.0	10,983	0.8	178.1	0.9	56	72	148
Kentucky	1,421	0.9	12,068	0.8	173.8	0.9	86	63	143
Louisiana	1,547	1.0	11,744	0.8	149.1	0.8	36	20	68
Maine	645	0.4	4,946	0.3	67.4	0.3	88	40	132
Maryland	2,303	1.5	17,719	1.2	292.2	1.5	98	86	219
Massachusetts	3,099	2.0	15,519	1.1	291.3	1.5	131	45	167
Michigan	3,664	2.4	30,418	2.1	437.4	2.2	97	68	146
Minnesota	2,287	1.5	17,073	1.2	264.8	1.3	30	32	116
Mississippi	1,488	1.0	13,542	0.9	168.3	0.8	36	3	63
Missouri	2,426	1.6	16,402	1.1	221.8	1.1	59	48	118
Montana	627	0.4	3,120	0.2	39.8	0.2	83	57	125
Nebraska	1,359	0.9	9,179	0.6	141.9	0.7	119	110	190

Nevada	622	0.4	5,005	0.4	74.9	0.4	114	119	184
New Hampshire	596	0.4	3,064	0.2	44.6	0.2	112	82	172
New Jersey	4,412	2.9	23,371	1.6	427.3	2.2	72	49	192
New Mexico	804	0.5	11,216	0.8	108.5	0.5	54	90	122
New York	6,778	4.4	39,368	2.8	686.0	3.5	46	48	157
North Carolina	3,063	2.0	30,367	2.1	426.5	2.2	102	79	216
North Dakota	496	0.3	2,126	0.1	30.3	0.2	-5	17	70
Ohio	4,219	2.8	34,330	2.4	495.7	2.5	45	80	158
Oklahoma	1,254	0.8	8,832	0.6	120.6	0.6	74	45	108
Oregon	2,271	1.5	29,270	2.0	359.9	1.8	65	96	208
Pennsylvania	4,459	2.9	37,808	2.6	603.5	3.0	55	56	159
Rhode Island	630	0.4	2,593	0.2	44.1	0.2	48	48	161
South Carolina	1,451	0.9	12,956	0.9	162.5	0.8	78	50	166
South Dakota	467	0.3	1,883	0.1	29.5	0.1	94	44	117
Tennessee	1,539	1.0	12,469	0.9	162.7	0.8	61	61	172
Texas	11,293	7.4	87,297	6.1	1,100.9	5.6	148	63	120
Utah	536	0.3	4,312	0.3	53.0	0.3	74	102	170
Vermont	416	0.3	2,374	0.2	36.6	0.2	120	100	239
Virginia	2,603	1.7	23,249	1.6	344.3	1.7	111	123	286
Washington	11,701	7.6	74,415	5.2	733.1	3.7	357	115	164
West Virginia	411	0.3	3,099	0.2	37.9	0.2	49	40	133
Wisconsin	2,101	1.4	18,642	1.3	311.5	1.6	44	61	150
Wyoming	380	0.2	2,082	0.1	26.4	0.1	90	28	78
United States	**153,105**	**100.0**	**1,429,905**	**100.0**	**19,822.1**	**100.0**	**60**	**39**	**115**

Source: ES-202 Employment and Wages Program; Bureau of Labor Statistics, U.S. Department of Labor.

* Agriculture includes crops (SIC–01), livestock (02) and agricultural services (07).

Most states have Federal Unemployment Tax Act coverage of farm employers; farm employers who pay quarterly cash wages of $20,000 or more or employ 10 or more workers for at least one day in each of 20 different weeks must cover their farmworkers (20-10 rule). CA, ME, MN, RI, TX, and WA have more inclusive farmworker coverage.

Percent changes may be exaggerated in states, including WA and TX, that extended coverage beyond the Federal minimum during the 1980s.

Employers are farms or their subunits which employ 50 or more workers.

Avg. Ann. Employment is the employer-reported count of all workers on the payroll for the pay period which includes the twelfth day of the month summed over 12 months and divided by 12.

Appendix Table 8. Crop Employment and Wages Covered by Unemployment Insurance in 1990

State	Employers		Employment		Wages		Percent Change: 1980-90		
	Number	% of U.S.	Avg. Annual	% of U.S.	($millions)	% of U.S.	Employers	Employment	Wages
Alabama	173	0.4	3,460	0.6	42.9	0.6	48	46	116
Alaska	9	0.0	99	0.0	1.3	0.0	42	25	63
Arizona	603	1.3	11,630	2.1	153.1	2.2	36	-5	27
Arkansas	504	1.1	3,448	0.6	44.8	0.6	17	-24	-3
California	17,731	39.2	200,800	36.6	2,707.8	38.5	-4	-13	36
Colorado	328	0.7	4,642	0.8	66.8	0.9	51	25	111
Connecticut	141	0.3	4,248	0.8	68.8	1.0	25	-15	115
Delaware	51	0.1	787	0.1	15.0	0.2	18	25	154
Florida	2,438	5.4	63,702	11.6	818.7	11.6	66	18	91
Georgia	375	0.8	7,245	1.3	80.4	1.1	62	35	142
Hawaii	179	0.4	7,901	1.4	158.7	2.3	143	-7	36
Idaho	475	1.1	8,067	1.5	98.8	1.4	94	88	162
Illinois	603	1.3	9,889	1.8	159.5	2.3	71	44	118
Indiana	353	0.8	6,509	1.2	88.7	1.3	40	29	73
Iowa	258	0.6	2,265	0.4	34.0	0.5	43	-40	-24
Kansas	238	0.5	1,799	0.3	29.7	0.4	113	70	165
Kentucky	127	0.3	1,268	0.2	14.5	0.2	38	13	56
Louisiana	443	1.0	4,637	0.8	54.9	0.8	15	-3	34
Maine	119	0.3	1,542	0.3	19.6	0.3	115	40	158
Maryland	189	0.4	2,399	0.4	35.0	0.5	46	33	138
Massachusetts	252	0.6	3,002	0.5	53.9	0.8	73	-2	110
Michigan	728	1.6	11,818	2.2	141.1	2.0	61	36	85
Minnesota	484	1.1	4,527	0.8	64.1	0.9	7	14	79
Mississippi	471	1.0	5,094	0.9	52.8	0.7	0	-34	-6
Missouri	415	0.9	3,651	0.7	46.1	0.7	32	-2	45
Montana	102	0.2	613	0.1	8.3	0.1	161	68	177
Nebraska	220	0.5	1,439	0.3	21.5	0.3	778	554	552
Nevada	41	0.1	543	0.1	8.8	0.1	426	925	1000

234

State									
New Hampshire	39	0.1	539	0.1	7.4	0.1	38	1	68
New Jersey	719	1.6	6,840	1.2	114.3	1.6	16	7	115
New Mexico	149	0.3	2,961	0.5	33.7	0.5	109	47	103
New York	846	1.9	9,998	1.8	146.6	2.1	31	16	82
North Carolina	560	1.2	7,989	1.5	84.9	1.2	89	50	148
North Dakota	233	0.5	1,152	0.2	16.4	0.2	-25	9	62
Ohio	464	1.0	9,392	1.7	120.1	1.7	16	43	100
Oklahoma	104	0.2	2,125	0.4	29.4	0.4	79	46	118
Oregon	904	2.0	19,193	3.5	223.4	3.2	92	70	172
Pennsylvania	504	1.1	12,149	2.2	179.2	2.5	23	34	110
Rhode Island	84	0.2	613	0.1	11.2	0.2	4	8	75
South Carolina	211	0.5	4,117	0.8	41.7	0.6	41	6	73
South Dakota	39	0.1	260	0.0	4.2	0.1	726	519	1300
Tennessee	198	0.4	3,342	0.6	45.6	0.6	10	56	187
Texas	3,130	6.9	24,281	4.4	263.3	3.7	521	81	125
Utah	50	0.1	1,026	0.2	11.7	0.2	92	66	121
Vermont	43	0.1	364	0.1	4.4	0.1	82	29	100
Virginia	297	0.7	4,583	0.8	55.4	0.8	115	76	184
Washington	8,003	17.7	53,766	9.8	456.5	6.5	797	122	167
West Virginia	61	0.1	581	0.1	5.5	0.1	69	-34	-4
Wisconsin	448	1.0	5,903	1.1	92.8	1.3	36	33	123
Wyoming	42	0.1	249	0.0	3.3	0.0	115	83	200
United States	45,185	100.0	548,503	100.0	7,041.7	100.0	45	12	67

Source: ES-202 Emplyment and Wages Program; Bureau of Labor Statistics, U.S. Department of Labor.

Most states have Federal Unemployment Tax Act coverage of farm employers; farm employers who pay quarterly cash wages of $20,000 or more or employ 10 or more workers for at least one day in each of 20 different weeks must cover their farmworkers (20-10 rule). CA, ME, MN, RI, TX, and WA have more inclusive farmworker coverage.

Percent changes may be exagerrated in states, including WA and TX, that extended coverage beyond the Federal minimum during the 1980s.

Employers are farms or their subunits which employ 50 or more workers.

Avg. Ann. Employment is the employer-reported count of all workers on the payroll for the pay period which includes the twelfth day of the month summed over 12 months and divided by 12.

Appendix Table 9. Livestock Employment and Wages Covered by Unemployment Insurance in 1990

State	Employers Number	Employers % of U.S.	Employment Avg. Annual	Employment % of U.S.	Wages ($millions)	Wages % of U.S.	Percent Change: 1980-90 Employers	Percent Change: 1980-90 Employment	Percent Change: 1980-90 Wages
Alabama	117	0.8	2,971	1.9	51.7	2.1	38	-2	79
Alaska	4	0.0	24	0.0	0.5	0.0			
Arizona	221	1.5	2,676	1.7	44.8	1.8	80	12	70
Arkansas	144	1.0	4,725	3.1	76.4	3.1	29	55	166
California	4,146	27.3	28,165	18.4	483.4	19.4	-14	-12	45
Colorado	304	2.0	3,679	2.4	60.4	2.4	93	68	116
Connecticut	68	0.4	1,081	0.7	24.5	1.0	51	43	214
Delaware	21	0.1	720	0.5	13.8	0.6	73	8	94
Florida	587	3.9	7,506	4.9	117.4	4.7	52	1	58
Georgia	196	1.3	3,937	2.6	68.6	2.7	19	2	91
Hawaii	59	0.4	969	0.6	18.5	0.7	6	0	75
Idaho	208	1.4	2,657	1.7	43.5	1.7	669	518	846
Illinois	181	1.2	1,323	0.9	19.9	0.8	91	27	105
Indiana	210	1.4	3,311	2.2	58.6	2.3	146	76	190
Iowa	249	1.6	2,559	1.7	40.9	1.6	99	74	157
Kansas	289	1.9	3,804	2.5	76.1	3.1	52	54	128
Kentucky	121	0.8	1,976	1.3	32.3	1.3	27	-15	30
Louisiana	76	0.5	721	0.5	10.7	0.4	42	-4	53
Maine	31	0.2	708	0.5	10.2	0.4	18	-43	-10
Maryland	104	0.7	1,276	0.8	20.4	0.8	78	-13	26
Massachusetts	66	0.4	559	0.4	9.1	0.4	38	-51	-28
Michigan	163	1.1	1,669	1.1	25.2	1.0	276	91	227
Minnesota	291	1.9	3,742	2.4	59.5	2.4	10	43	144
Mississippi	143	0.9	4,076	2.7	61.5	2.5	145	86	200
Missouri	200	1.3	2,347	1.5	35.1	1.4	116	56	172
Montana	181	1.2	1,482	1.0	19.6	0.8	121	47	106
Nebraska	341	2.2	3,787	2.5	69.3	2.8	123	95	175

Nevada	89	0.6	1,069	0.7	14.5	0.6	59	-6	19
New Hampshire	0	0.0	0	0.0	0.0	0.0	-100	-100	-100
New Jersey	87	0.6	651	0.4	12.2	0.5	37	18	139
North Carolina	224	1.5	6,990	4.6	129.6	5.2	81	50	189
North Dakota	0	0.0	0	0.0	0.0	0.0	-100	-100	-100
Ohio	155	1.0	2,761	1.8	39.7	1.6	45	113	254
Oklahoma	186	1.2	1,783	1.2	28.3	1.1	68	42	101
Oregon	139	0.9	1,737	1.1	25.6	1.0			
Pennsylvania	195	1.3	3,445	2.2	53.2	2.1	37	22	119
Rhode Island	40	0.3	132	0.1	1.4	0.1	14	-45	-13
South Carolina	68	0.4	1,592	1.0	23.0	0.9	33	44	171
South Dakota	101	0.7	654	0.4	12.0	0.5	105	28	122
Tennessee	58	0.4	407	0.3	5.6	0.2	3	-37	12
Texas	2,547	16.8	19,052	12.4	282.7	11.3	303	65	135
Utah	74	0.5	949	0.6	14.8	0.6	128	142	270
Vermont	41	0.3	395	0.3	6.5	0.3	768	913	1525
Virginia	260	1.7	3,021	2.0	48.2	1.9	127	78	205
Washington	1,234	8.1	5,415	3.5	78.4	3.1	1022	125	189
West Virginia	42	0.3	272	0.2	4.1	0.2	151	192	356
Wisconsin	207	1.4	2,092	1.4	31.9	1.3	168	78	170
Wyoming	127	0.8	1,153	0.8	15.8	0.6	81	7	52
United States	**15,180**	**100.0**	**153,155**	**100.0**	**2,494.9**	**100.0**	**54**	**26**	**101**

Source: ES-202 Emplyment and Wages Program; Bureau of Labor Statistics, U.S. Department of Labor.

Most states have Federal Unemployment Tax Act coverage of farm employers; farm employers who pay quarterly cash wages of $20,000 or more or employ 10 or more workers for at least one day in each of 20 different weeks must cover their farmworkers (20-10 rule). CA, ME, MN, RI, TX, and WA have more inclusive farmworker coverage.

Percent changes may be exagerated in states, including WA and TX, that extended coverage beyond the Federal minimum during the 1980s.

Employers are farms or their subunits which employ 50 or more workers.

Avg. Ann. Employment is the employer-reported count of all workers on the payroll for the pay period which includes the twelfth day of the month summed over 12 months and divided by 12.

Appendix Table 10. Top Ten States in Employers, Average Employment, and Wages Paid for All Agriculture*

State	Employers Number	% of U.S.	Employment Avg. Ann.	% of U.S.	Wages ($millions)	% of U.S.
California	34,576	23	437,507	31	5,899.2	30
Washington	11,701	8	141,845	10	1,838.4	9
Texas	11,293	7	87,297	6	1,100.9	6
Florida	10,601	7	74,415	5	733.1	4
New York	6,778	4	39,368	3	686.0	3
Pennsylvania	4,459	3	37,808	3	607.3	3
New Jersey	4,412	3	36,537	3	603.5	3
Ohio	4,219	3	34,330	2	495.7	3
Illinois	4,009	3	33,504	2	437.4	2
Michigan	3,664	2	30,418	2	427.3	2
Top Ten States	95,711	63	953,029	67	12,828.8	65
United States	153,105	100	1,429,905	100	19,822.1	100

Source: ES-202 Emplyment and Wages Program; Bureau % of Labor Statistics, U.S. Department % of Labor.

* Agriculture includes crops (SIC=01), livestock (02) and agricultural services (07).

Most states have Federal Unemployment Tax Act coverage % of farm employers; farm employers who pay quarterly cash wages % of $20,000 or more or employ 10 or more workers for at least one day in each % of 20 different weeks must cover their farmworkers (20-10 rule). CA, ME, MN, RI, TX, and WA have more inclusive farmworker coverage.

Percent changes may be exagerrated in states, including WA and TX, that extended coverage beyond the Federal minimum during the 1980s.

Employers are farms or their subunits which employ 50 or more workers.

Avg. Ann. Employment is the employer-reported count % of all workers on the payroll for the pay period which includes the twelfth day % of the month summed over 12 months and divided by 12.

Appendix Table 11. Top Ten Employment States in the FVH, FAS, and FLC Industries

State	Fruits, Vegetables and Horticulture* Av. Annual	% of U.S.	Farm Agricultural Services** Av. Annual	% of U.S.	Farm Labor Contractors Av. Annual	% of U.S.
California	153,119	38	130,891	56	74,811	66
Florida	58,216	14	34,077	15	20,598	18
Washington	43,219	11	13,649	6	6,053	5
Oregon	14,013	3	9,622	4	5,192	5
Texas	12,583	3	5,502	2	1,111	1
Pennsylvania	11,673	3	3,829	2	491	0
New York	9,470	2	3,159	1	423	0
Michigan	9,172	2	2,784	1	251	0
Ohio	7,605	2	2,684	1	180	0
Arizona	7,006	2	2,476	1	149	0
United States	402,802	100	232,538	100	113,964	100

Source: ES-202 Emplyment and Wages Program; Bureau % of Labor Statistics, U.S. Department % of Labor.

*FVH includes fruits and nuts (016), vegetables and melons (017) and horticultural specialties (018).

**FAS consists % of 071, 072 and 076 (which includes 0761, farm labor contractors).

Most states have Federal Unemployment Tax Act coverage % of farm employers; farm employers who pay quarterly cash wages % of $20,000 or more or employ 10 or more workers for at least one day in each % of 20 different weeks must cover their farmworkers (20-10 rule). CA, ME, MN, RI, TX, and WA have more inclusive farmworker coverage.

Avg. Ann. Employment is the employer-reported count % of all workers on the payroll for the pay period which includes the twelfth day % of the month summed over 12 months and divided by 12.

Appendix Table 12. Ranking % of Regions by Wage and Number % of Workers: 1990

Region	Wage Hourly	% of U.S.	Region	Workers % (Thousands)	% of U.S.
Hawaii	8.61	156	California	168	20
California	6.34	115	Lake	77	9
Florida	6.00	109	Southern Plains	60	7
Pacific	5.95	108	Cornbelt I	55	7
Cornbelt I	5.59	101	Florida	50	6
Northeast I	5.51	100	Southeast	48	6
Northern Plains	5.41	98	Northeast I	47	6
Northeast II	5.34	97	Appalachian I	45	5
Mountain III	5.32	96	Delta	43	5
Mountain II	5.29	96	Appalachian II	40	5
Lake	5.19	94	Northeast II	40	5
Cornbelt II	5.07	92	Northern Plains	38	5
Southern Plains	4.97	90	Cornbelt II	37	4
Mountain I	4.85	88	Mountain I	25	3
Appalachian II	4.84	88	Mountain II	24	3
Southeast	4.82	87	Mountain III	17	2
Appalachian I	4.69	85	Hawaii	10	1
Delta	4.66	84	Pacific	6	1
United States	5.52	100	United States	828	100

Source: NASS Farm Labor Reports, U.S. Department % of Agriculture.

Average annual wage for all hired workers and number % of hired workers.

The National Agricultural Statistics Service collects these data primarily through phone interviews with farm employers; Mail surveys are also used. Historically, these surveys have been conducted four times a year: in January, April, July and October.

Appalachian I: NC, VA; Appalachian II: KY, TN, WV; Cornbelt I: IL, IN, OH; Cornbelt II: IA, MO; Delta: AR, LA, MS; Lake: MI, MN, WI; Mountain I: ID, MT, WY; Mountain II: CO, NV, UT; Mountain III; AZ, NM; Northeast I: CT, ME, MA, NH, NY, RI, VT; Northeast II: DE, MD, NJ, PA; Northern Plains: KS, NE, ND, SD; Pacific: OR, WA; Southeast: AL, GA, SC; Southern Plains: OK, TX.

Appendix Table 13. Average Annual Hourly Wages Reported in Farm Labor: 1975-1990

Region	Nominal Wage Rates				Percent Changes			
	1975	1980	1985	1990	1975-80	1980-85	1985-90	1975-90
Appalachian I	2.06	3.19	3.98	4.69	55	25	18	128
Appalachian II	2.02	3.23	3.90	4.84	60	21	24	140
California	2.91	4.51	5.47	6.34	55	21	16	118
Cornbelt I	2.37	3.63	4.39	5.59	53	21	27	136
Cornbelt II	2.35	3.63	4.21	5.07	54	16	20	116
Delta	2.22	3.38	3.88	4.66	52	15	20	110
Florida	2.82	4.19	4.99	6.00	49	19	20	113
Hawaii		5.58	7.45	8.61		34	16	
Lake	2.31	3.31	3.95	5.19	43	19	31	125
Mountain I	2.39	3.37	3.87	4.85	41	15	25	103
Mountain II	2.46	3.55	4.35	5.29	44	23	22	115
Mountain III	2.37	3.43	4.53	5.32	45	32	17	124
Northeast I		3.09	4.01	5.51		30	37	
Northeast II	2.39	3.52	4.12	5.34	47	17	30	123
Northern Plains	2.40	3.59	4.51	5.41	50	26	20	125
Pacific	2.78	4.07	4.83	5.95	46	19	23	114
Southeast	2.13	3.18	3.54	4.82	49	11	36	126
Southern Plains	2.19	3.35	4.46	4.97	53	33	11	127
United States	**2.43**	**3.66**	**4.42**	**5.52**	**51**	**21**	**25**	**127**

Source: NASS Farm Labor Reports, U.S. Department % of Agriculture.

Wage rates are average annual wages for all hired farmworkers.

The National Agricultural Statistics Service collects these data primarily through phone interviews with farm employers. Some mailed surveys are also used. Historically, these surveys have been conducted four times a year: in January, April, July and October. The hourly wage rate is calculated by dividing the sum % of the reported wages paid during the pay period % of the survey week (that which includes the 12th % of the month) by the sum % of the reported hours worked during that pay period.

Appalachian I: NC, VA; Appalachian II: KY, TN, WV; Cornbelt I: IL, IN, OH; Cornbelt II: IA, MO; Delta: AR, LA, MS; Lake: MI, MN, WI; Mountain I: ID, MT, WY; Mountain II: CO, NV, UT; Mountain III; AZ, NM; Northeast I: CT, ME, MA, NH, NY, RI, VT; Northeast II: DE, MD, NJ, PA; Northern Plains: KS, NE, ND, SD; Pacific: OR, WA; Southeast: AL, GA, SC; Southern Plains: OK, TX.

Appendix Table 14. Number % of Hired Workers Reported in Farm Labor: 1975-1990

Region	% of Workers (Thousands)				Percent Changes			
	1975	1980	1985	1990	1975-80	1980-85	1985-90	1975-90
Appalachian I	86	81	173	45	-6	113	-74	-48
Appalachian II	46	55	36	40	21	-34	11	-12
California	240	197	177	168	-18	-10	-5	-30
Cornbelt I	90	90	77	55	0	-15	-29	-39
Cornbelt II	64	63	49	37	-2	-22	-26	-43
Delta	81	82	58	43	1	-29	-27	-48
Florida	59	68	55	50	15	-19	-9	-15
Hawaii		11	11	10	NA	0	-11	NA
Lake	118		197	77	-100	NA	-61	-35
Mountain I	35	35	28	25	0	-19	-13	-30
Mountain II	26	28	24	24	5	-14	3	-7
Mountain III	23	23	19	17	-1	-18	-12	-28
Northeast I	65	75	60	47	15	-20	-22	-28
Northeast II	58	60	51	40	5	-15	-21	-30
Northern Plains	63	67	57	38	7	-15	-34	-40
Pacific	3	4	5	6	46	17	25	114
Southeast	97	87	66	48	-10	-24	-27	-50
Southern Plains	98	94	82	60	-4	-13	-27	-39
United States	1,317	1,303	1,224	828	-1	-6	-32	-37

Source: NASS Farm Labor Reports, U.S. Department % of Agriculture.

The National Agricultural Statistics Service collects these data primarily through phone interviews with farm employers. Some mailed surveys are also used. Historically, these surveys have been conducted four times a year: in January, April, July and October.

Appalachian I: NC, VA; Appalachian II: KY, TN, WV; Cornbelt I: IL, IN, OH; Cornbelt II: IA, MO; Delta: AR, LA, MS; Lake: MI, MN, WI; Mountain I: ID, MT, WY; Mountain II: CO, NV, UT; Mountain III: AZ, NM; Northeast I: CT, ME, MA, NH, NY, RI, VT; Northeast II: DE, MD, NJ, PA; Northern Plains: KS, NE, ND, SD; Pacific: OR, WA; Southeast: AL, GA, SC; Southern Plains: OK, TX.

NOTES

[1]For other reviews of the farm labor data problem see Stan G. Daberkow & Leslie A. Whitener, Agricultural Labor Data Sources: An update (USDA, Economic Research Service, Agriculture Handbook 658, 1986) and James Holt et al., Toward Definition and Measurement of Farm Employment (USDA and AAEA workshop, Washington, DC, May 1977).

[2]There is a problem with COA county-level data in states such as California, which had 1,000 farms with more than one establishment or unit in the 1969 COA. The COA simply tabulates each sub-unit in the county in which it is located, thus understating the degree of concentration in agriculture. Thus, Newhall Land and Farming, an NYSE-listed company with operations in 5 California counties, appears in the COA as 5 operations, not the one farming company that it is.

The Census of Manufactures does not confuse companies and establishments because it reports companies and establishments separately. The fact that there are today more than 1,000 California farming companies that have more than one establishment or sub-unit emphasizes just how much California farms are factories-in-the-fields. I am indebted to Don Villarejo for clarifying county-level COA reporting procedures.

[3] Holt et. al. supra note 1, at II.7.

[4] In the 1982 COA, about 879,000 farms reported hiring almost 5 million workers, and labor expenditures were $8.4 billion. Some of these farms also obtain labor through contractors or crew leaders; about 140,000 farms reported expenditures of $1.1 billion for contract labor. Finally, some farm operators rely on custom machine operators to plow or combine, and their expenditures for such custom services include the machine rental fee and the wages of the operator. About 787,000 farms reported expenditures of $2 billion for such custom services in 1982.

[5]Fieldworkers are employed primarily in crops.

[6]The $5.1 billion in FVH labor expenditures includes $4.1 billion for workers hired directly and $1 billion for contract labor. Horticultural specialties (nursery and greenhouse products) accounted for 40 percent of FVH direct hire expenditures, and fruits and nuts accounted for almost 60 percent of FVH contract labor expenditures. Nearly half of all FVH labor expenditures were paid by California farmers.

[7]Once a farmer or landowner is located, QALS interviewers reportedly follow-up to locate all of the farm workers employed on this land, whether employed by the farmer directly or by a FLC.

[8]Note that the $5.11 figure is not the wage that either irrigators or harvest workers are paid, even though employers who apply for H-2A workers are required to develop detailed job descriptions for harvesters or irrigators and offer the higher of the AEWR, prevailing, or minimum wage for such a detailed job.

[9]The QALS began publishing national estimates of agricultural service employment in 1977.

[10]Whitener, 1983.

[11] In 1987 1,600 persons who did farm work: were interviewed. About 500 worked on livestock farms, 800 on grain farms, and 400 in fruit and vegetable agriculture. Only 196 of the fruit and vegetable workers did 25 or more days of farm work in 1987. Critics suggest that the CPS misses Hispanic migrants employed in fruit and vegetable agriculture who are out of the country in December.

[12]Unpublished data indicate that 49 percent of the sample migrants did farm work in the state where they were enumerated in 1981, 78 percent in 1983, and 56 percent in 1985.

[13]Both QALS and NAWS interviews have problems getting farm employers to provide wage and employment data (QALS) or provide their list of workers (NAWS). In some cases, 30 to 50 percent contacted employers.

[14]The top 20 states included 101 percent of the currently children because a currently migratory child can be reported by 2 or more states.

Appendix D. Glossary of Acronyms

ACUS   Administrative Conference of the United States
AFDC   Aid to Families with Dependent Children
AFOP   Association of Farmworker Opportunity Programs
ALRA   Agricultural Labor Relations Act
ALRB   Agricultural Labor Relations Board
BLS   Bureau of Labor Statistics
CAMP   College Assistance Migrant Program
CAW   Commission on Agricultural Workers
CES   Current Employment Statistics
CETA   Comprehensive Employment and Training Act
CHC   Community Health Center
COA   Census of Agriculture
COE   Certificate of Eligibility
COP   Census of Population
CPS   Current Population Survey
CRESS   Clearinghouse on Rural Education and Small Schools
DOL   Department of Labor
ECIA   Education Consolidation and Improvement Act
EPA   Environmental Protection Agency
ERIC   Educational Resources Information Center
ETA   Employment and Training Administration
FLA   Farm Labor Alliance
FLC   Farm Labor Contractor
FLOC   Farm Labor Organizing Committee
FLSA   Fair Labor Standards Act
FTE   Full-Time Equivalent
FVH   Fruit, Vegetable and Horticultural Commodities
GAO   General Accounting Office
HEP   High School Equivalency Program
HEW   Department of Health, Education and Welfare
HFWF   Hired Farm Working Force
HHS   Department of Health and Human Services
IRCA   Immigration Reform and Control Act
JTPA   Job Training Partnership Act
LEA   Local Education Agency
LSC   Legal Services Corporation
ME   Migrant Education
MEP   Migrant Education Program
MH   Migrant Health
MHC   Migrant Health Center
MHS   Migrant Head Start
MSFW   Migrant and Seasonal Farm Workers
MSRTS   Migrant Student Record Transfer System
NACHC   National Association of Community Health Centers
NACMH   National Advisory Council on Migrant Health

NAFTA	North American Free Trade Agreement
NASDME	National Association of State Directors of Migrant Education
NAWS	National Agricultural Workers Survey
NCME	National Commission on Migrant Education
NLRA	National Labor Relations Act
NMRP	National Migrant Resource Program
OEO	Office of Economic Opportunity
OMB	Office of Management and Budget
OSTP	Office of Science and Technology Policy
PCC	Program Coordination Center
PIC	Private Industry Council
QALS	Quarterly Agricultural Labor Survey
RAW	Replenishment Agricultural Worker
SAS	Seasonal Agricultural Services
SAW	Special Agricultural Worker
SEA	State Education Agency
SIC	Standard Industrial Classification
SOC	Standard Occupational Classification
UI	Unemployment Insurance
USDA	United States Department of Agriculture
WC	Workers' Compensation
WIC	Supplemental Food Program for Women Infants and Children

Index